AF784343

Intelligent Microgrid Management and EV Control Under Uncertainties in Smart Grid

Ran Wang · Ping Wang · Gaoxi Xiao

Intelligent Microgrid Management and EV Control Under Uncertainties in Smart Grid

Springer

Ran Wang
College of Computer Science and Technology
Nanjing University of Aeronautics and Astronautics
Nanjing, Jiangsu
China

Gaoxi Xiao
School of Electrical and Electronic Engineering
Nanyang Technological University
Singapore
Singapore

Ping Wang
School of Computer Science and Engineering
Nanyang Technological University
Singapore
Singapore

ISBN 978-981-10-4249-2 ISBN 978-981-10-4250-8 (eBook)
https://doi.org/10.1007/978-981-10-4250-8

Library of Congress Control Number: 2017958017

© Springer Nature Singapore Pte Ltd. 2018
This work is subject to copyright. All rights are reserved by the Publisher, whether the whole or part of the material is concerned, specifically the rights of translation, reprinting, reuse of illustrations, recitation, broadcasting, reproduction on microfilms or in any other physical way, and transmission or information storage and retrieval, electronic adaptation, computer software, or by similar or dissimilar methodology now known or hereafter developed.
The use of general descriptive names, registered names, trademarks, service marks, etc. in this publication does not imply, even in the absence of a specific statement, that such names are exempt from the relevant protective laws and regulations and therefore free for general use.
The publisher, the authors and the editors are safe to assume that the advice and information in this book are believed to be true and accurate at the date of publication. Neither the publisher nor the authors or the editors give a warranty, express or implied, with respect to the material contained herein or for any errors or omissions that may have been made. The publisher remains neutral with regard to jurisdictional claims in published maps and institutional affiliations.

Printed on acid-free paper

This Springer imprint is published by Springer Nature
The registered company is Springer Nature Singapore Pte Ltd.
The registered company address is: 152 Beach Road, #21-01/04 Gateway East, Singapore 189721, Singapore

Acknowledgements

This monograph would not have been possible without a lot of people who helped me and changed my life profoundly during my research career.

First and foremost, I would like to express my deepest gratitude to my Ph.D. supervisor, Dr. Ping Wang and Dr. Gaoxi Xiao. Their infectious enthusiasm, unlimited patience, and inspiring guidance have been the major driving force through my research career at NTU. Under their supervision and training, I have developed skills in mathematical analysis, technical writing, communication skills, and leadership, which will be invaluable to my future career.

Beyond my supervisors, I would like to thank Dr. Dusit Niyato, who have provided many insightful comments and contributed significantly to my research work. In addition, I also would like to thank the book examiners for their comments and suggestions to improve the quality of this monograph.

Moreover, I would like to thank the students and staffs in Multimedia Graduate Laboratory at the School of Computer Science and Engineering in Nanyang Technological University. The laboratory peers, including Weiwen Zhang, Yichao Jin, Yuan Yuan, Guanyu Gao, Yiqun Yang, Shimin Gong, Yifan Li, and Xiao Lu, have not only provided me with their valuable suggestions for my research but also enriched my research life at NTU with enjoyable experiences. Besides, I would also like to thank the laboratory staffs for their warmhearted assistants throughout my research activities.

Finally, I cannot end without giving my special thanks to my parents, my sister, and my fiancee. My family have always provided me with support and stood by me through ups and downs. Thanks very much for their unselfish love and unconditioned supports.

Contents

List of Figures

List of Tables

Abstract

A modern power grid needs to become smarter in order to provide an affordable, reliable, and sustainable supply of electricity. For these reasons, a smart grid is necessary to manage and control the increasingly complex future grid. Certain smart grid elements including renewable energy, storage, microgrid, consumer choice, and smart appliances like electric vehicles increase uncertainty in both supply and demand of electric power.

In this book, we investigate the intelligent control of two important components in smart grid, namely microgrids (MGs) and electric vehicles (EVs). We focus on developing theoretical frameworks and proposing corresponding algorithms, to optimally schedule virtualized elements (e.g., conventional generators' output, electricity imported, EVs' charging rates, and customers' energy demand) under different uncertainties (e.g., renewable energy generation uncertainty, energy demand uncertainty, EVs' pattern uncertainty, electricity price uncertainty), so that the total cost of operating the microgrid or the EV charging system can be minimized and the systems maintain stabilized. First, we consider power demand and supply management problem in microgrid with uncertain renewable energy integration. To model the randomness of renewable energy generation, a novel uncertainty model is developed. An optimization problem is then formulated to determine the optimal power consumption and generation scheduling for minimizing the fuel cost. We propose a two-stage optimization approach to solve the problem. The second case considers energy generation scheduling in the microgrid. For this case, we develop robust optimization-based techniques to tackle the uncertainties from net demand, heat demand, and electricity prices. It is shown that our energy generation scheduling strategy performs well which can effectively reduce the system expenditure. Based on the framework in the second case, we further investigate the energy generation scheduling problem in microgrids involving temporal-correlated renewable energy. Under such case, chance constraint approximations and robust optimization approaches based on a Chebyshev inequality framework are developed to first transform and then solve the scheduling problem. Experimental results show that temporal-correlation information of the renewable energy within a proper time span can effectively reduce the conservativeness of the problem solution.

Next, we consider charging scheduling of a large number of EVs at a charging station which is equipped with renewable energy generation devices. Stimulated by the fact that in practical scenario, EV arrival and renewable energy may not follow any determinate process yet obtaining some statistical information of future EVs' arrivals (departures) is possible, we propose a novel two-stage EV charging mechanism to minimize the cost and efficiently utilize renewable energy. Several uncertain quantities such as the arrival and departure times of the EVs, their charging requirements, and available renewable energy are taken into account. Finally in the last case, we develop a hybrid centralized–decentralized (HCD) EV charging scheme which offers flexible charging choices for customers. In this charging scheme, EV owners can either assign the charging tasks to system controller or individually choose the charging profiles based on their own preferences. In addition, the stochastic characteristics of EVs such as the arrival and departure times and charging demands are taken into account.

The aforementioned microgrid management policies and EV charging schemes can effectively reduce the operational cost of the systems. The proposed approaches and obtained results may provide guidelines to improve the efficiency of the smart grid operation and provide useful insights helping system operators develop rational investment strategies.

Chapter 1
Introduction

In this chapter, we first present the background regarding the components of power system, transformation from traditional grid to smart grid and smart grid's unique characteristics. Then, we specifically discuss two important components in smart grid, namely microgrids (MGs) and electric vehicles (EVs). Intelligent control of these two components under uncertainties is our objective and research interests. Finally, the organization of this book is illustrated.

1.1 Background

1.1.1 Electrical Power Systems

Electrical power system is an interconnected assemblage of elements and networks in order to generate, transfer, and consume the electrical energy. Power system components are divided into three general categories: generators, transmission and distribution network, and consumers.

- **Generators**: A generator, which is the main component of each power plant, converts different types of energy into electrical power. Most of the generators burn fossil fuels such as natural gas, oil, and coal to produce electricity, and some of them utilize nuclear energy, but the usage of renewable sources such as wind, solar, geothermal heat, biomass, and hydroelectric energy has been increasing in recent years as well.
- **Transmission and distribution network**: Transmission system carries electrical energy from suppliers in generation side to electrical substations in demand side. Using the transmission network, electricity is transferred at high voltage (110 kV and higher) in order to decrease the power loss during the transmission of the power. Overhead power lines are usually utilized as transmission networks. Underground

© Springer Nature Singapore Pte Ltd. 2018

R. Wang et al., *Intelligent Microgrid Management and EV Control Under Uncertainties in Smart Grid*, https://doi.org/10.1007/978-981-10-4250-8_1

power lines are used just in the city or sensitive areas because they have higher cost and operational restrictions. At final step, the distribution network receives electrical power in substations from transmission lines. After reducing the level of voltage (less than 50 kV) by substations, distribution system delivers electricity to consumers in demand side.

- **Consumers**: The last component of power systems is consumers. Consumers or loads receive electrical power from distribution network as end users. The size of loads varies from small household appliances to huge industrial machinery.

1.1.2 Transition to a Smart Grid

The utility industry across the world is trying to address numerous challenges, including generation diversification, optimal deployment of expensive assets, demand response, energy conservation, and reduction of the industry overall carbon footprint. It is evident that such critical issues cannot be addressed within the confines of the existing electricity grid.

The existing electricity grid is unidirectional in nature. Its topology is shown in Fig. 1.1. In this hierarchical configuration, a failure in any component is transferred to other components in the chain and may result in poor power quality, such as power cuts or even blackouts. This system converts only one-third of fuel energy into electricity, without recovering the waste heat. Almost 8% of its output is lost along its transmission lines, while 20% of its generation capacity exists to meet peak demand, which happens only 5% of the time [1].

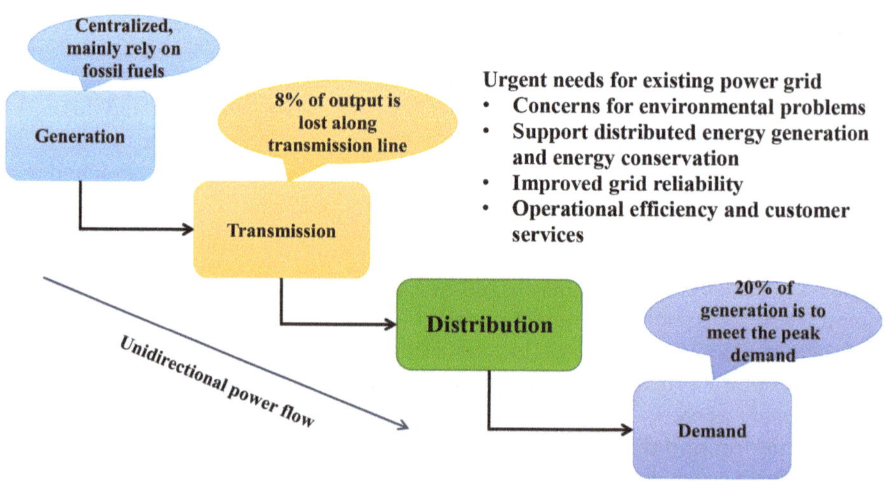

Fig. 1.1 Unidirectional power flow from generation side to demand side

Furthermore, continuous increase in the electricity consumption around the world places considerable stress on aging power system. It is projected that electricity usage in the USA will increase from 3873 TWh in 2008 to 5021 TWh in 2035. Summer peak demand in the USA is expected to increase by 40% from 2008 to 2030 as well. Environmental pollution and global warming due to the use of fossil fuels for electricity generation and depletion of fossil fuel reserves have already raised serious concerns about sustainable operation of power systems in the future.

It is widely believed that the present electric power system is undergoing a profound change driven by these urgent needs, including the concerns on environmental problems and the need for energy conservation. We need improved grid reliability while dealing with an aging infrastructure. We also need better operational efficiency and customer services. For this reason, the current electric power grid will need to be transformed "smarter." This transformation is to meet the environmental target, to support distributed energy generation and storage, and to satisfy the urgent requirement for demand response (DR) and renewable energy integration.

The smart grid (SG), also called smart electrical/power grid, intelligent grid, is expected to address the major shortcomings of the existing grid. Regarded as the next-generation power grid, smart grid uses two-way flows of electricity and information to create a widely distributed and automatic energy delivery network [2]. By utilizing modern information technologies, the smart grid is capable of delivering power in more efficient ways and responding to wide-ranging conditions and events [3]. Broadly stated, the smart grid could respond to events that occur anywhere in the grid, such as power generation, transmission, distribution, and consumption, and adopt the corresponding strategies. For instance, once a medium voltage transformer failure event occurs in the distribution grid, the smart grid may automatically change the power flow and recover the power delivery service. Let us consider another example of demand profile shaping. Since lowering peak demand and smoothing demand profile reduces overall plant and capital cost requirements, in the peak period the electric utility can use real-time pricing to convince some users to reduce their power demands, so that the total demand profile full of peaks can be shaped to a nicely smoothed demand profile.

More specifically, the smart grid can be regarded as an electric system that uses information, two-way, cyber-secure communication technologies, and computation intelligence in an integrated fashion across electricity generation, transmission, substation, distribution, and consumption to achieve a system that is clean, safe, secure, reliable, resilient, efficient, and sustainable [4]. This description covers the entire spectrum of the energy system from the generation to the end points of consumption of the electricity. The ultimate smart grid is a vision. Given the vast landscape of the smart grid research, different researchers may express different visions for smart grid due to different focus and perspectives. An example of the general operation architecture of smart grid is shown in Fig. 1.2, where the system consists of four parts, namely power generation and control system, power transmission and control system, residential distribution system, and communication network system.Table 1.1 depicts the salient features of the smart grid in comparison with the existing grid.

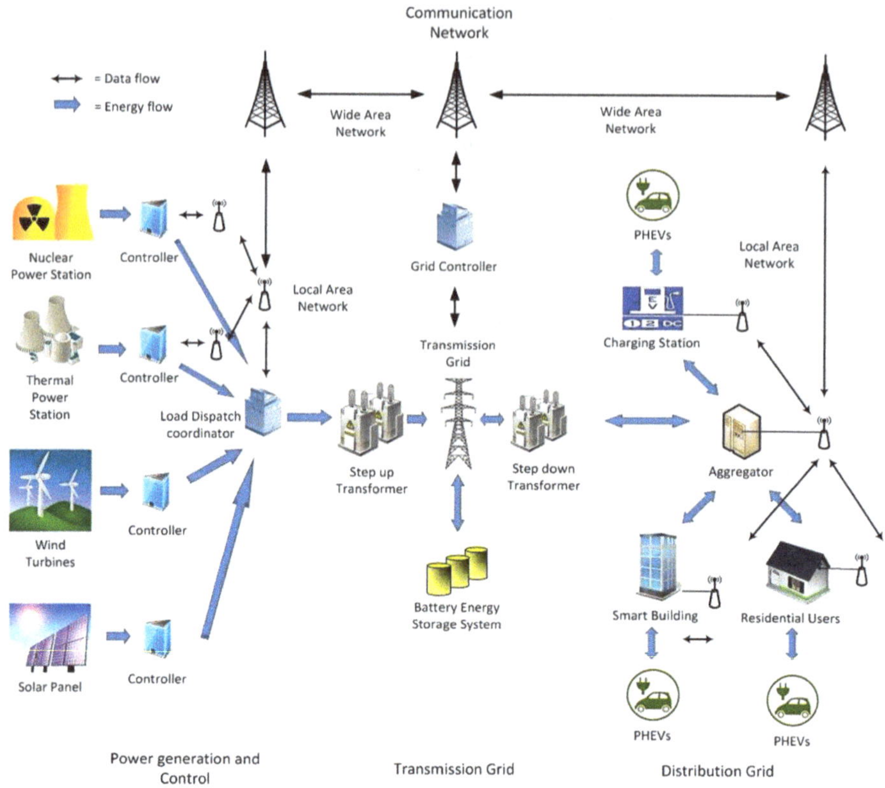

Fig. 1.2 An example of the operation framework of smart grid

Table 1.1 A brief comparison between the existing grid and the smart grid

Existing grid	Smart grid
Electromechanical	Digital
One-way communication	Two-way communication
Centralized generation	Distributed generation
Few sensors	Sensors throughout
Manual monitoring	Self-monitoring
Manual restoration	Self-healing
Failures and blackouts	Adaptive and islanding
Limited control	Pervasive control
Few customer choices	Many customer choices

1.1.3 Microgrids (MGs) and Electric Vehicles (EVs)

To allow pervasive control and monitoring, the smart grid is emerging as a convergence of information technology and communication technology with power system engineering. In smart grid, two-way flows of electricity and information are supported, which lay the foundation for realizing various functions and management objectives. The grid will keep becoming smarter with the development of new management applications and services that can leverage the technology and capability upgrades enabled by this advanced infrastructure.

Within the framework of smart grid, many new components, which are difficult and even infeasible to be integrated into conventional power grids, become possible, easy, or even indispensable. Among them, two important components in the future smart grid are microgrids (MGs) and electric vehicles (EVs).

- The growth and evolution of the smart grid is expected to come with the plug-and-play integration of the basic structures called microgrids, which represents the future paradigm of power systems. Specifically, microgrids are small-scale low-voltage power supply networks designed to supply electrical load for a small community such as a university campus, a commercial area, and a trading estate. Microgrids can autonomously coordinate local generations and demands in a dynamic manner. It can operate in either grid-connected mode or islanded mode [5]. There have been worldwide deployments of pilot microgrids, e.g., in US, Germany, Greece, and Japan [6]. The power generators or microsources employed in microgrids are usually renewable or non-conventional distributed energy resources. While incorporating such renewable resources shall bring great environmental benefits, it imposes new challenges as well: Different from that in the traditional power systems with conventional controllable electric generators, generation scheduling in microgrids with fluctuant, climate-dependent renewable energy sources has to cope with the non-trivial uncertainties. In addition, small-scale demands in micgorids are also hard to predict and the real-time pricing in electricity market yields another uncertainty dimension. Considering these challenges, intelligent energy management of microgrids to achieve a robust and cost-effective goal still remains an open issue.
- Electric vehicles (EVs) are gaining attentions as a cleaner alternative to fossil fuel vehicles. To encourage the purchase of EVs, government of different countries including Australia, Canada, China, Europe Union, and USA subsidize or finance the customers and implement many actions such as tax exemption, transit and parking facility constructions [7]. On the one hand, the higher penetration of EVs not only makes the transportation sector less carbon-intensive, but also plays an important role in improving the overall stability and efficiency of the power system by playing an active role in the demand-side management (DSM), spinning reserve services, and reducing the impact of uncertainties brought by renewable energy integration. On the other hand, it will create significant new load on the distribution network which brings up multiple technical issues, such as voltage deviations, transformers and line saturations, increase of electrical losses, making

the integration of EVs into the power grid very challenging. Therefore, integrating massive EVs into the power grid requires the charging to be coordinated to realize the actual benefits and eliminate the harmful impacts.

1.2 Research Focus

The objective of this book is to design sustainable, reliable, and cost-effective control schemes to intelligently manage the operations of two important components of future smart grid, i.e., microgrids and EVs. Specifically, we aim to develop theoretical frameworks and propose corresponding algorithms, to optimally schedule virtualized resources or elements (e.g., conventional generators' output, electricity imported, EVs' charging rates, and customers' energy demand) in various combinations under different uncertainties (e.g., renewable energy generation uncertainty, energy demand uncertainty, EVs' pattern uncertainty, electricity price uncertainty), so that the total cost of operating the microgrid or the EV charging system can be minimized and the corresponding systems maintain stabilized. In particular, we focus on the following scenarios.

A. Energy Management in Microgrid

- We first consider the demand and supply management problem in microgrids considering the uncertainty of renewable energy generation. We focus on the control of home appliances and distributed generators to achieve a cost-effective scheduling.
- Then, we consider the energy generation scheduling problem in a microgrid system, in which uncertainties from combined heat and power (CHP) generators, renewable energy resources, and electricity prices are taken into account. We aim at effectively scheduling different energy sources to achieve a reliable and cost-effective control.
- Based on the framework in the second scenario, we further investigate the energy generation scheduling in a microgrid system involving temporal-correlated renewable energy. Our motivation is to study how the temporal-correlation information impacts the scheduling performance.

B. Electric Vehicle Charging

- Furthermore, we consider charging scheduling of a large number of EVs at a charging station which is equipped with renewable energy generation devices. The uncertainties from renewable energy, EVs' arrival and departure times, and EVs' charging amount should be all taken into account.
- Finally, we investigate a hybrid centralized–decentralized EV charging scheme which offers flexible charging choices for customers. Model predictive control-based technique and game theory concepts are adopted to formulate the problem. Uncertainties of EVs' arrival and departure times and their energy demands should be specifically considered.

1.3 Organization of the Chapters

Figure 1.3 shows the main content and organization of the book. Specifically, we can divide the main content into two parts, including energy management in microgrids and EVs' charging control. The first half of the book (Chaps. 3 and 4) considers demand and supply management and energy generation scheduling in microgrids. The second half of the book develops two EV charging schemes under two different scenarios.

In summary, the rest of this book is organized as follows:

- **Chapter** 2 introduces related works on the energy management problems in microgrids and EV charging strategies.
- **Chapter** 3 develops a demand and supply energy management scheme which aims at minimizing the monetary cost of operating the microgrid system by strategically determining the operation processes of home appliances and distributed generators.
- **Chapter** 4 formulates a cost minimization problem to intelligently schedule energy generations for microgrids equipped with unstable renewable sources and combined heat and power (CHP) generators. We aim at effectively scheduling different energy sources.
- **Chapter** 5 studies the energy generation scheduling problem in microgrids involving unstable temporal-correlated renewable energy. We focus on investigating how the temporal-correlated information of renewable energy impacts the scheduling performance of a microgrid.
- **Chapter** 6 investigates the cost-effective scheduling approach of EV charging at a renewable energy-aided charging station. In addition, a fast charging rate

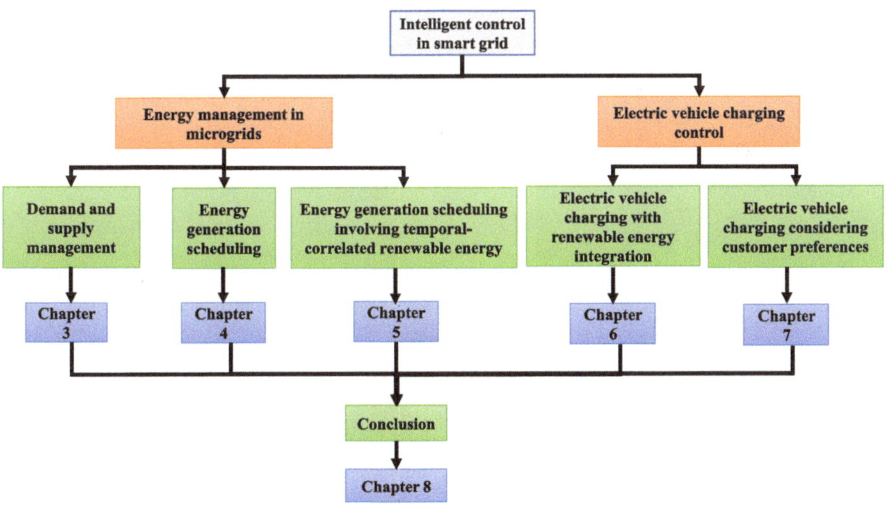

Fig. 1.3 Main content and organization of the book

compression algorithm is developed which tremendously reduces the complexity of the problem solving.

- **Chapter** 7 investigates the coordination of EVs' charging at a charging park considering the EV owners' various charging preferences. A hybrid centralized–decentralized (HCD) charging mechanism is designed to effectively schedule the charging of EVs and stimulate the EV owners' energy demands.
- **Chapter** 8 summarizes the book and discusses the further work for intelligent control of microgrids and EVs.

References

1. H. Farhangi, The path of the smart grid. IEEE Power Energy Mag. **8**(1), 18–28 (2010)
2. X. Fang, S. Misra, G. Xue, D. Yang, Smart grid: the new and improved power grid: A survey. IEEE Commun. Surv. Tutor. **14**(4), 944–980 (2012)
3. A. Ipakchi, F. Albuyeh, Grid of the future. IEEE Power Energy Mag. **7**(2), 52–62 (2009)
4. H. Gharavi, R. Ghafurian, Smart grid: The electric energy system of the future. Proc. IEEE **99**(6), 917–921 (2011)
5. S. Chowdhury, S. Chowdhury, P. Crossley, Microgrids and Active Distribution Networks. The Institution of Engineering and Technology (2009)
6. M. Barnes, J. Kondoh, H. Asano, J. Oyarzabal, G. Ventakaramanan, R. Lasseter, N. Hatziargyriou, T. Green, Real-world microgrids-an overview. in IEEE International Conference on System of Systems Engineering, pp. 1–8, IEEE (2007)
7. J. García-Villalobos, I. Zamora, J. San Martín, F. Asensio, V. Aperribay, Plug-in electric vehicles in electric distribution networks: A review of smart charging approaches. Renew. Sustain. Energy Rev. **38**, 717–731 (2014)

Chapter 2
Literature Review

In this chapter, we provide a literature review of intelligent microgrid management and electric vehicle charging control with different decision objectives in different scenarios. We first give an overview of the energy management mechanisms in microgrids. We then review existing works concerning electric vehicle charging strategies. The limitations of previous literature and the advantages of our method over theirs are analyzed.

2.1 Energy Management in Microgrid

2.1.1 Supply and Demand Management

This problem can be viewed as containing two different parts. On the power supply side, we need to build a hierarchical demand control scheme so as to achieve the economic consumption scheduling and fulfill the requirements set by energy users; on the power demand side, there is a need to properly model the randomness of renewable energy generation, which may account for a significant portion of power supply in microgrids. Note that load balance constraints act as the connection between power consumption and generation.

Demand control techniques can be categorized into either price-based load control techniques, referred to as demand response methods, or direct load control, referred to as demand-side management. Under price-based load control scheme, users are encouraged to make energy consumption decisions individually according to the price information. Demand-side management strategies, however, are usually applied directly by a central controller and require consumer subscription to an economic incentive program. Some representative work has studied demand control techniques in residential microgrids. A recent paper [1] develops a real-time pricing

© Springer Nature Singapore Pte Ltd. 2018

R. Wang et al., *Intelligent Microgrid Management and EV Control Under Uncertainties in Smart Grid*, https://doi.org/10.1007/978-981-10-4250-8_2

scheme which aims at reducing the peak-to-average load ratio (PAR) through demand response management in smart grid systems. A two-stage optimization problem is proposed and solved. Fathi et al. develop a stochastic model for scheduling in a local area network with the objective of cost minimization and PAR minimization [2]. The work in [3] presents a linear programming formulation for minimizing the energy cost through direct load control. In [4], a robust optimization approach is presented to adjust the hourly load level of a given consumer in response to hourly electricity prices. The uncertainties of renewable energies, however, are not considered in these studies. As such, the control schemes may not be readily optimal and applicable to the microgrid scenario where renewable energies constitute a significant portion of power resources.

There also exist some studies considering renewable energy uncertainties when scheduling the energy generation. Such work can be categorized into two groups: the stochastic-based approaches and the robust optimization-based approaches. For instance, Wang et al. define stochastic upper and lower supply curves to capture a broad range of fluctuations in the power system, where energy generated by each power source is modeled as stochastic arrivals in the queuing model [5]. In [6], scenario-based stochastic operation management methods are developed to tackle the fluctuant demands and renewable energies using the probability distribution function (PDF) of each uncertain variable. Hidden Markov models have also been adopted to characterize renewable energy generation [7–9]. Stimulated by observations that in practical scenarios, obtaining an accurate distribution function could be computationally costly and renewable energy may not follow Markov process or any simple distributions, robust optimization has recently received growing attention as a modeling framework for optimization under uncertainty. Instead of assuming explicit probability distribution, robust optimization confines the renewable generation in a pre-defined uncertainty set containing the worst-case scenario. For example, Zhang et al. consider a distributed economic dispatch problem for microgrid with high penetration of renewable energies [10]. The intrinsically stochastic properties of renewable energy sources are captured by a polyhedral uncertainty set with deterministic lower and upper bounds. Similar methods for modeling renewable energies can also be found in other recent work [11, 12]. The main topics concerning supply and demand management in microgrids are illustrated in Fig. 2.1.

Different from the existing work, our approach in Chap. 3 jointly considers power demand and supply management. Rather than assuming there is available knowledge of the specific distribution of renewable energy generation, the proposed approach describes the underlying uncertainty in a more detailed yet flexible manner. It allows more information of renewable energy generation to be effectively incorporated into the uncertainty model when such information is available.

2.1.2 Energy Generation Scheduling

Energy generation scheduling is the process of effectively scheduling different energy sources (local generators, central grid, renewable energy generations, etc.) to meet

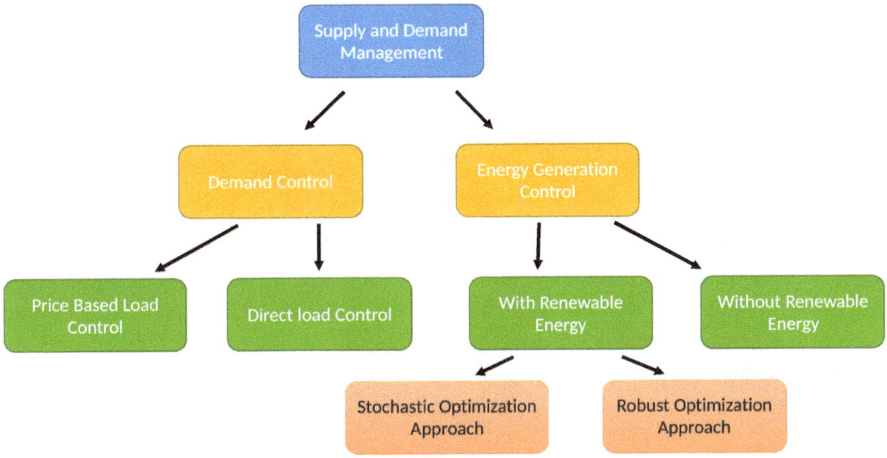

Fig. 2.1 Main topics concerning supply and demand management in microgrids

the energy requests at a minimum cost subject to various physical constraints of the power systems. It is a classic problem in electricity system which is composed of two aspects, namely unit commitment (UC) [13] and economic dispatch (ED) [14]. The UC problem involves determining the start-up and shut-down schedules of generator units to be used to meet forecast demand over a short time in future. It is a complex optimization problem with both integer and continuous variables and has been shown to be NP-Complete in general. The basic UC methods reported in the literature include priority listing [15], dynamic programming [16], Lagrangian relaxation [17], integer programming [18, 19]. After UC problem has determined the start-up and shut-down schedules, the ED problem seeks to find the optimal allocation of electric power outputs from various available generators without alternating their on/off status. Readers can refer to comprehensive surveys on UC [20] and ED [21] for more details.

Conventional energy generation scheduling is typically conducted 24 h in advance (day-ahead) and based on the fact that the system load can be forecast with reasonably good accuracy one day in advance. In microgrids, however, this is no longer the case due to the fact that accurate predictions of small-scale electricity and heat demands, renewable energy supplies, and electricity market prices are very difficult, as we stated earlier. Some recent literature has investigated energy generation scheduling of microgrids [22–26]. In [22], a multi-objective optimization of economic load dispatch for a microgrid is investigated using evolutionary computation. The paper aims at minimizing the emission of the thermal generators and minimizing the total operating cost. In [23], a generalized formulation for intelligent energy management of microgrid is proposed using artificial intelligence techniques jointly with linear-programming-based multi-objective optimization. Similarly in [24], an intelligent energy management system is proposed for optimal operation of a CHP-based microgrid over a 24-h time interval. Authors of [25, 26] also propose different energy management strategies based on different assumptions. The limitation of these results, however, is that they all assume that the energy demands and supplies are known ahead of time, which is rarely the case in practice.

There also exist some studies considering demand and supply uncertainties when scheduling the energy generation. These works can be categorized into two groups: the stochastic optimization-based approaches [6, 27–31] and robust optimization-based approaches [10, 32–35]. In [27], the author develops a solution method for scheduling units of a power-generating system to produce electricity by taking into consideration the stochastic nature of the hourly load and its correlation structure. In [28], a stochastic model for the long-term solution of security-constrained unit commitment is proposed. A more complicated scenario can be found in [6], in which an efficient stochastic framework is developed to investigate the effect of uncertainty on the operation management of microgrids. The proposed stochastic framework would consider the uncertainties of load forecast error, wind turbine generation, photovoltaic generation, and market price concurrently. Paper [29] examines the impact of the stochastic nature of wind on planning and dispatch of a system. Similarly, authors of [30] compare stochastic and reserve methods and evaluate the benefits of a combined approach for the efficient management of uncertainty in the unit commitment problem. In [31], a two-stage stochastic objective function aiming at minimizing the expected operational cost is implemented. The stochastic optimization approaches[1] explicitly incorporate a probability distribution function of the uncertainty, and they often rely on enumerating discrete scenarios of the uncertainty realizations. Such approaches mainly have two practical limitations. First, it may be difficult and costly to obtain an accurate probability distribution of uncertainty. Second, the solution only provides probabilistic guarantees to the system reliability. To obtain enough high guarantee requires a huge number of samples, which poses substantial computational challenges.

In recent literature, robust optimization has received growing attention as a modeling framework for optimization under uncertainty. In [32], a two-stage adaptive robust unit commitment model is proposed for the security-constrained unit commitment problem in the presence of nodal net injection uncertainty. In [33], a robust optimization approach is proposed to accommodate wind output uncertainty, with the objective of providing a robust unit commitment schedule for the thermal generators in the day-ahead market. In [10], a power scheduling approach is proposed based on robust optimization to address the intrinsically stochastic availability of renewable energy sources. Papers [34, 35] also present robust optimization-based approach for optimal microgrid management considering wind power or energy consumption uncertainties. Instead of postulating explicit probability distribution,

[1] As an example, consider two-stage linear programs. Here the decision maker takes some action in the first stage, after which a random event occurs affecting the outcome of the first-stage decision. A recourse decision can then be made in the second stage that compensates for any bad effects that might have been experienced as a result of the first-stage decision. The optimal policy from such a model is a single first-stage policy and a collection of recourse decisions (a decision rule) defining which second-stage action should be taken in response to each random outcome.

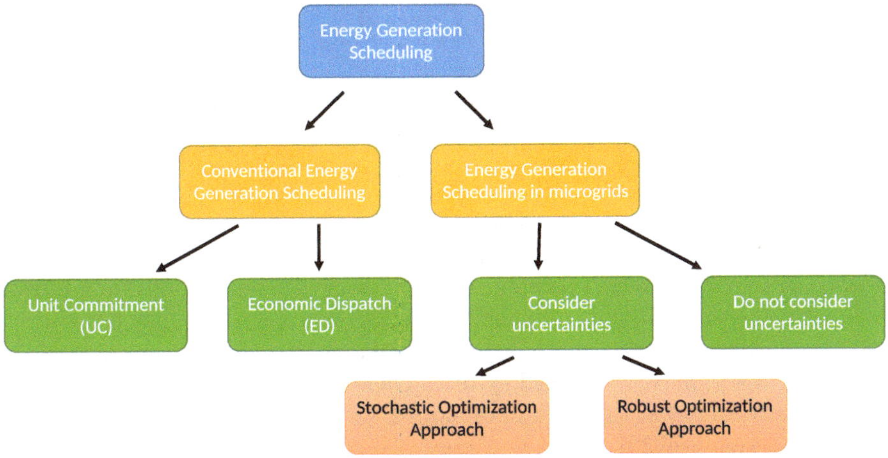

Fig. 2.2 Main topics concerning energy generation scheduling in electrical grids

robust optimization confines the random variable in a pre-defined uncertainty set containing the worst-case scenario. For instance, in [10–12, 32–35], uncertainties in price prediction or renewable energy generation are presented as interval values with deterministic lower and upper bounds, and the framework developed in [36, 37] is incorporated to solve the problem. Without requiring explicit probability distribution, uncertainty can be characterized more flexibly. In addition, the conservativeness of the solution can easily be controlled, and the problem is always computationally tractable both practically and theoretically even for large-scale problems. The main topics concerning energy generation scheduling in electrical grids are illustrated in Fig. 2.2.

In our study (Chap. 4), robust optimization concept is also applied to tackle the uncertainties in energy generation scheduling problem of microgrids. Different from the previous robust optimization works [10–12, 32–35] which confine the uncertainty within a lower and upper bounds, in our work, we propose a new uncertainty model to characterize the renewable energy and user demand uncertainties, which can provide more statistical details in describing the underlying uncertainty. Moreover, the proposed uncertainty model is also flexible enough that we can incorporate more information into the uncertainty model when such information is available. Whereas in Chap. 5, we further focus on investigating how the temporal-correlation information of the renewable energy impacts the scheduling performance bounds based on the framework in Chap. 4. To the best of our knowledge, we are the first to do such evaluations.

2.2 Electric Vehicle Charging Control

The existing EVs' charging scheduling mechanisms can be roughly classified into two categories: centralized charging strategies and decentralized charging strategies. The main idea of centralized control is utilizing centralized infrastructure to collect information from all EVs and centrally optimize EVs' charging considering the grid technical constraints. In such a strategy, the master controller makes decisions about the rate and time of charging EVs to get the optimal solution. References [38–41] develop various centralized charging strategies with different optimization objectives, including saving system cost, minimizing CO_2 emission, reducing power loss, adjusting power frequency, and satisfying EV owners. Either optimization methods or heuristic algorithms are adopted by researchers to solve such problems. In [42], a hierarchical control scheme is proposed for EVs' charging station loads in a distribution network while minimizing energy cost and abiding by substation supply constraints. The scheduling is based on the forecasted load information. Reference [43] proposes a dynamic programming (DP)-based optimization method of charging an EV fleet modeled as a single, so-called aggregate battery. In these papers, the dynamics of the EVs' arriving/departing times and charging patterns are not considered. Recent literatures [44–47] all adopt receding horizon-control-based techniques to tackle the uncertainties in the dynamic charging systems. References [48–50] develop online algorithms for coordinating the EVs' charging to save the system cost and lessen the EVs' harmful impacts on the distribution network. Jin et al. [51] study EV charging scheduling problems from a customer's perspective by jointly considering the aggregator's revenue and customers' demands and costs. Paper [52] studies risk-aware day-ahead scheduling and real-time dispatch for plug-in EVs, aiming to jointly optimize the EV charging cost and minimizing the risk of the load mismatch between the forecast and the actual EV loads. Different from previous papers, both static and dynamic charging scenarios are considered in [51, 52]. Though the centralized charging strategy is straightforward, the size of the centralized optimization increases with the number of EVs. Accurate information collection from a large number of EVs may also impose a challenge. Designing an effective centralized EV charging strategy therefore remains as a difficult problem.

In contrast, the vehicle owners can directly control their EVs' charging patterns employing the decentralized charging strategies [53–68]. Gan et al. [53] propose a decentralized algorithm to schedule EV charging to fill the electric load valley. This charging control strategy iteratively solves an optimal control problem in which the charging rate of each vehicle can vary continuously within its upper and lower bounds. In each iteration, each EV updates its own charging profile according to the control signal broadcast by the utility, and the utility company alters the control signal to guide their updates. In [54–60], various decentralized charging frameworks to coordinate charging demand of EVs are implemented based on game theory concepts. In [61], a decentralized online valley filling algorithm for EV charging is proposed. An optimal power flow (OPF) framework is adopted to model the network constraint that arises from charging EVs at different locations. Similarly, decentralized EV charging schemes with valley filling objective can be found in [62, 63]. Considering the selfish nature of people, authors of [64] define some weighting factors in the objective function of EV charging management problem aiming at modeling users' convenience in the presented optimization procedure. Xi et al. [65] study a decentralized price-based EV charging control. They study a pricing scheme that conveys price and quantity information to the load aggregator and compare it to a simpler price-only scheme. In [66], a novel online coordination method for the charging of plug-in EVs in smart distribution networks is proposed. An innovative parking lot prediction unit is developed adopting $M/G/\infty$ queuing model. In [67], the authors formulate the EV charging problem as a convex optimization problem and then propose a decentralized water-filling-based algorithm to solve it. A receding horizon approach (similar to [44–46]) is utilized to handle the random arrival of EVs and the inaccuracy of the forecast non-EV load. Although the decentralized charging strategy offers more ownership authority to EV owners, it may not ensure optimality in the charging of EVs and causes security concerns of the power grid [38, 54, 68].

In the above-mentioned literature, the charging energy is supplied purely from power grid, largely generated by conventional units. The main goal of introducing EVs, namely reducing the pollution and greenhouse gas of transportation sector, is consequently greatly abated, as the pollution is transferred from vehicle itself to conventional energy units. Renewable energy should play a role as significantly as possible to achieve the real environmental advantage. Renewable-energy-based EV charging hence becomes a practical and critical problem.

Though the topic has not been well investigated in the literature, a few related works can still be found dealing with the charging scheduling of EVs with renewable energy integration. Moeini et al. [69] propose a charging management framework considering multiple criteria including total loss of distribution networks, rescheduling cost, and wind energy utilization. In [69], it is assumed that the energy demand of EVs is known by the controller. In [70], a price-incentive model is utilized to generate the management strategy to coordinate the charging of EVs and battery swapping station (BSS). While in [71], the mathematical models are built for both smart charging and V2G operation with distribution grid constraints. Authors in [70, 71] both assume that the EVs are static and always available to be charged/discharged. In [72], a stochastic optimization algorithm is presented to coordinate charging of electric-drive vehicles (EDVs) in order to maximize the utilization of renewable energy in transportation. Due to the stochastic nature of transportation patterns, the Monte Carlo simulation is applied to model uncertainties presented by numerous scenarios. In [73], the charging problem is formulated as a stochastic semi-Markov decision process with the objective of maximizing the energy utilization. In recent work [74], the uncertainties of the EV arrival and renewable energy are described as independent Markov processes. In [75, 76], the authors tackle the EV charging scheduling problem adopting Lyapunov optimization techniques, such that statistics of the underlying processes does not need to be known in prior. The main topics concerning EV charging scheduling are illustrated in Fig. 2.3.

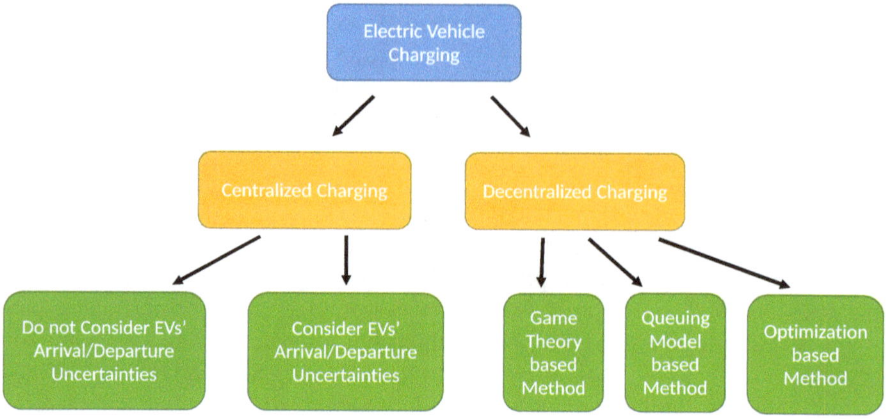

Fig. 2.3 Main topics concerning EV charging scheduling

Compared with what has been proposed in the past, our EV charging mechanism proposed in Chap. 6 mainly shows the following several advantages: (1) Renewable energies can be effectively utilized by the EVs; (2) compared with the online scheduling schemes, the proposed mechanism incorporates useful estimated information day-ahead to help reduce the uncertainties in the real-time scheduling stage; (3) compared with the offline scheduling schemes, our mechanism is fairly flexible such that it can effectively respond to real-time incidents; (4) a fast computing algorithm is designed which can easily tackle a large number of EVs; i.e., one weakness of the centralized charging strategies is overcome. Whereas in Chap. 7, compared with previous studies, the proposed hybrid centralized–decentralized (HCD) EV charging scheme offers flexible charging choices for customers, where EV owners can either assign the charging tasks to system controller or individually choose the charging profiles based on their own preferences. The stochastic characteristics of EVs such as the arrival/departure times and charging demands are all taken into account. Moreover, the communication burden between EVs and the system controller is low, and the proposed charging scheme is robust to poor communication channels.

References

1. L.P. Qian, Y.J.A. Zhang, J. Huang, Y. Wu, Demand response management via real-time electricity price control in smart grids. IEEE J. Sel. Areas Commun. **31**(7), 1268–1280 (2013)
2. M. Fathi, H. Bevrani, Adaptive energy consumption scheduling for connected microgrids under demand uncertainty. IEEE Trans. Power Deliv. **28**(3), 1576–1583 (2013)
3. A.-H. Mohsenian-Rad, A. Leon-Garcia, Optimal residential load control with price prediction in real-time electricity pricing environments. IEEE Trans. Smart Grid **1**(2), 120–133 (2010)
4. A.J. Conejo, J.M. Morales, L. Baringo, Real-time demand response model. IEEE Trans. Smart Grid **1**(3), 236–242 (2010)

5. K. Wang, F. Ciucu, C. Lin, S.H. Low, A stochastic power network calculus for integrating renewable energy sources into the power grid. IEEE J. Sel. Areas Commun. **30**(6), 1037–1048 (2012)
6. S. Mohammadi, S. Soleymani, B. Mozafari, Scenario-based stochastic operation management of microgrid including wind, photovoltaic, micro-turbine, fuel cell and energy storage devices. Int. J. Electr. Power Energy Syst. **54**, 525–535 (2014)
7. S. Bu, F.R. Yu, P.X. Liu, Stochastic unit commitment in smart grid communications, in *2011 IEEE Conference on Computer Communications Workshops* IEEE (2011), pp. 307–312
8. S. Bu, F.R. Yu, P.X. Liu, P. Zhang, Distributed scheduling in smart grid communications with dynamic power demands and intermittent renewable energy resources, in *2011 IEEE International Conference on Communications Workshops* IEEE (2011), pp. 1–5
9. T. Soubdhan, R. Emilion, R. Calif, Classification of daily solar radiation distributions using a mixture of dirichlet distributions. Sol. Energy **83**(7), 1056–1063 (2009)
10. Y. Zhang, N. Gatsis, G.B. Giannakis, Robust energy management for microgrids with high-penetration renewables. IEEE Trans. Sustain. Energy **4**(4), 944–953 (2013)
11. R. Jabr, Robust transmission network expansion planning with uncertain renewable generation and loads. IEEE Trans. Power Syst. **28**(4), 4558–4567 (2013)
12. S.-J. Kim, G.B. Giannakis, Scalable and robust demand response with mixed-integer constraints. IEEE Trans. Smart Grid **4**(4), 2089–2099 (2013)
13. S.A. Kazarlis, A. Bakirtzis, V. Petridis, A genetic algorithm solution to the unit commitment problem. IEEE Trans. Power Syst. **11**(1), 83–92 (1996)
14. Z.-L. Gaing, Particle swarm optimization to solving the economic dispatch considering the generator constraints. IEEE Trans. Power Syst. **18**(3), 1187–1195 (2003)
15. R. Johnson, H. Happ, W. Wright, Large scale hydro-thermal unit commitment-method and results. IEEE Trans. Power Appar. Syst. **3**, 1373–1384 (1971)
16. C. Pang, G.B. Sheblé, F. Albuyeh, Evaluation of dynamic programming based methods and multiple area representation for thermal unit commitments. IEEE Trans. Power Appar. Syst. **3**, 1212–1218 (1981)
17. F. Zhuang, F.D. Galiana, Towards a more rigorous and practical unit commitment by lagrangian relaxation. IEEE Trans. Power Syst. **3**(2), 763–773 (1988)
18. T.S. Dillon, K.W. Edwin, H.-D. Kochs, R. Taud, Integer programming approach to the problem of optimal unit commitment with probabilistic reserve determination. IEEE Trans. Power Appar. Syst. **6**, 2154–2166 (1978)
19. M. Carrión, J.M. Arroyo, A computationally efficient mixed-integer linear formulation for the thermal unit commitment problem. IEEE Trans. Power Syst. **21**(3), 1371–1378 (2006)
20. N.P. Padhy, Unit commitment-a bibliographical survey. IEEE Trans. Power Syst. **19**(2), 1196–1205 (2004)
21. B.H. Chowdhury, S. Rahman, A review of recent advances in economic dispatch. IEEE Trans. Power Syst. **5**(4), 1248–1259 (1990)
22. J. Xu, S. Tan, S.K. Panda, Optimization of economic load dispatch for a microgrid using evolutionary computation, in *IEEE Annual Conference of Industrial Electronics Society* IEEE (2011), pp. 3192–3197
23. A. Chaouachi, R. Kamel, R. Andoulsi, K. Nagasaka, Multiobjective intelligent energy management for a microgrid. IEEE Trans. Ind. Inform. **60**(4), 1688–1699 (2013)
24. M. Motevasel, A.R. Seifi, T. Niknam, Multi-objective energy management of CHP (combined heat and power)-based micro-grid. Energy **51**, 123–136 (2013)
25. X. Wu, X. Wang, Z. Bie, Optimal generation scheduling of a microgrid, in *International Conference and Exhibition on Innovative Smart Grid Technologies* IEEE (2012), pp. 1–7
26. A. Parisio, L. Glielmo, A mixed integer linear formulation for microgrid economic scheduling, in *IEEE International Conference on Smart Grid Communications* IEEE (2011), pp. 505–510
27. U.A. Ozturk, M. Mazumdar, B.A. Norman, A solution to the stochastic unit commitment problem using chance constrained programming. IEEE Trans. Power Syst. **19**(3), 1589–1598 (2004)

28. L. Wu, M. Shahidehpour, T. Li, Stochastic security-constrained unit commitment. IEEE Trans. Power Syst. **22**(2), 800–811 (2007)
29. A. Tuohy, P. Meibom, E. Denny, M. O'Malley, Unit commitment for systems with significant wind penetration. IEEE Trans. Power Syst. **24**(2), 592–601 (2009)
30. P.A. Ruiz, C.R. Philbrick, E. Zak, K.W. Cheung, P.W. Sauer, Uncertainty management in the unit commitment problem. IEEE Trans. Power Syst. **24**(2), 642–651 (2009)
31. M. Mazidi, A. Zakariazadeh, S. Jadid, P. Siano, Integrated scheduling of renewable generation and demand response programs in a microgrid. Energy Convers. Manag. **86**, 1118–1127 (2014)
32. D. Bertsimas, E. Litvinov, X.A. Sun, J. Zhao, T. Zheng, Adaptive robust optimization for the security constrained unit commitment problem. IEEE Trans. Power Syst. **28**(1), 52–63 (2013)
33. R. Jiang, J. Wang, Y. Guan, Robust unit commitment with wind power and pumped storage hydro. IEEE Trans. Power Syst. **27**(2), 800–810 (2012)
34. R. Gupta, N.K. Gupta, A robust optimization based approach for microgrid operation in deregulated environment. Energy Convers. Manag. **93**, 121–131 (2015)
35. E. Kuznetsova, C. Ruiz, Y.-F. Li, E. Zio, Analysis of robust optimization for decentralized microgrid energy management under uncertainty. Int. J. Electr. Power Energy Syst. **64**, 815–832 (2015)
36. D. Bertsimas, M. Sim, Robust discrete optimization and network flows. Math. Progr. **98**(1), 49–71 (2003)
37. D. Bertsimas, M. Sim, The price of robustness. Oper. Res. **52**(1), 35–53 (2004)
38. M. Esmaili, A. Goldoust, Multi-objective optimal charging of plug-in electric vehicles in unbalanced distribution networks. Int. J. Electr. Power Energy Syst. **73**, 644–652 (2015)
39. A. Zakariazadeh, S. Jadid, P. Siano, Multi-objective scheduling of electric vehicles in smart distribution system. Energy Convers. Manag. **79**, 43–53 (2014)
40. M. Honarmand, A. Zakariazadeh, S. Jadid, Optimal scheduling of electric vehicles in an intelligent parking lot considering vehicle-to-grid concept and battery condition. Energy **65**, 572–579 (2014)
41. J. Yang, L. He, S. Fu, An improved pso-based charging strategy of electric vehicles in electrical distribution grid. Appl. Energy **128**, 82–92 (2014)
42. D.M. Anand, R.T. de Salis, Y. Cheng, J. Moyne, D.M. Tilbury, A hierarchical incentive arbitration scheme for coordinated pev charging stations. IEEE Trans. Smart Grid **6**(4), 1775–1784 (2015)
43. B. Škugor, J. Deur, Dynamic programming-based optimisation of charging an electric vehicle fleet system represented by an aggregate battery model. Energy **92**, 456–465 (2015)
44. W. Qi, Z. Xu, Z.-J.M. Shen, Z. Hu, Y. Song, Hierarchical coordinated control of plug-in electric vehicles charging in multifamily dwellings. IEEE Trans. Smart Grid **5**(3), 1465–1474 (2014)
45. M. Shaaban, A. Eajal, E. El-Saadany, Coordinated charging of plug-in hybrid electric vehicles in smart hybrid ac/dc distribution systems. Renew. Energy **82**, 92–99 (2015)
46. J. de Hoog, T. Alpcan, M. Brazil, D.A. Thomas, I. Mareels, Optimal charging of electric vehicles taking distribution network constraints into account. IEEE Trans. Power Syst. **30**(1), 365–375 (2015)
47. Y. He, B. Venkatesh, L. Guan, Optimal scheduling for charging and discharging of electric vehicles. IEEE Trans. Smart Grid **3**(3), 1095–1105 (2012)
48. W. Tang, S. Bi, Y.J. Zhang, Online coordinated charging decision algorithm for electric vehicles without future information. IEEE Trans. Smart Grid **5**(6), 2810–2824 (2014)
49. M. Moeini-Aghtaie, A. Abbaspour, M. Fotuhi-Firuzabad, Online multicriteria framework for charging management of phevs. IEEE Trans. Veh. Technol. **63**(7), 3028–3037 (2014)
50. L. Hua, J. Wang, C. Zhou, Adaptive electric vehicle charging coordination on distribution network. IEEE Trans. Smart Grid **5**(6), 2666–2675 (2014)
51. C. Jin, J. Tang, P. Ghosh, Optimizing electric vehicle charging: a customer's perspective. IEEE Trans. Veh. Technol. **62**(7), 2919–2927 (2013)
52. L. Yang, J. Zhang, H.V. Poor, Risk-aware day-ahead scheduling and real-time dispatch for electric vehicle charging. IEEE Trans. Smart Grid **5**(2), 693–702 (2014)

53. L. Gan, U. Topcu, S. Low, Optimal decentralized protocol for electric vehicle charging. IEEE Trans. Power Syst. **28**(2), 940–951 (2013)
54. Z. Ma, D.S. Callaway, I.A. Hiskens, Decentralized charging control of large populations of plug-in electric vehicles. IEEE Trans. Control Syst. Technol. **21**(1), 67–78 (2013)
55. R. Yu, J. Ding, W. Zhong, Y. Liu, S. Xie, Phev charging and discharging cooperation in v2g networks: a coalition game approach. IEEE Internet Things J. **1**(6), 578–589 (2014)
56. W. Lee, L. Xiang, R. Schober, V.W. Wong, Electric vehicle charging stations with renewable power generators: a game theoretical analysis. IEEE Trans. Smart Grid **6**(2), 608–617 (2015)
57. W. Tushar, W. Saad, H.V. Poor, D.B. Smith, Economics of electric vehicle charging: a game theoretic approach. IEEE Trans. Smart Grid **3**(4), 1767–1778 (2012)
58. E.L. Karfopoulos, N.D. Hatziargyriou, A multi-agent system for controlled charging of a large population of electric vehicles. IEEE Trans. Power Syst. **28**(2), 1196–1204 (2013)
59. A. Sheikhi, S. Bahrami, A. Ranjbar, H. Oraee, Strategic charging method for plugged in hybrid electric vehicles in smart grids; a game theoretic approach. Int. J. Electr. Power Energy Syst. **53**, 499–506 (2013)
60. S. Bahrami, M. Parniani, Game theoretic based charging strategy for plug-in hybrid electric vehicles. IEEE Trans. Smart Grid **5**(5), 2368–2375 (2014)
61. N. Chen, C.W. Tan, T.Q. Quek, Electric vehicle charging in smart grid: optimality and valley-filling algorithms. IEEE J. Sel. Top. Signal Process. **8**(6), 1073–1083 (2014)
62. K. Zhan, Z. Hu, Y. Song, N. Lu, Z. Xu, L. Jia, A probability transition matrix based decentralized electric vehicle charging method for load valley filling. Electr. Power Syst. Res. **125**, 1–7 (2015)
63. L. Zhang, F. Jabbari, T. Brown, S. Samuelsen, Coordinating plug-in electric vehicle charging with electric grid: Valley filling and target load following. J. Power Sour. **267**, 584–597 (2014)
64. C.-K. Wen, J.-C. Chen, J.-H. Teng, P. Ting, Decentralized plug-in electric vehicle charging selection algorithm in power systems. IEEE Trans. Smart Grid **3**(4), 1779–1789 (2012)
65. X. Xi, R. Sioshansi, Using price-based signals to control plug-in electric vehicle fleet charging. IEEE Trans. Smart Grid **5**(3), 1451–1464 (2014)
66. M.F. Shaaban, M. Ismail, E.F. El-Saadany, W. Zhuang, Real-time pev charging/discharging coordination in smart distribution systems. IEEE Trans. Smart Grid **5**(4), 1797–1807 (2014)
67. Y. Mou, H. Xing, Z. Lin, M. Fu, Decentralized optimal demand-side management for phev charging in a smart grid. IEEE Trans. Smart Grid **6**(2), 726–736 (2015)
68. S. Vandael, B. Claessens, M. Hommelberg, T. Holvoet, G. Deconinck, A scalable three-step approach for demand side management of plug-in hybrid vehicles. IEEE Trans. Smart Grid **4**(2), 720–728 (2013)
69. M. Moeini-Aghtaie, A. Abbaspour, M. Fotuhi-Firuzabad, Online multi-criteria framework for charging management of phevs. IEEE Trans. Veh. Technol. **63**(7), 3028–3037 (2014). future issue
70. M. Zhang, J. Chen, The energy management and optimized operation of electric vehicles based on microgrid. IEEE Trans. Power Deliv. **29**(3), 1427–1435 (2014)
71. S. Gao, K. Chau, C. Liu, D. Wu, C. Chan, Integrated energy management of plug-in electric vehicles in power grid with renewables. IEEE Trans. Veh. Technol. **63**(7), 3019–3027 (2014)
72. M. Pantoš, Stochastic optimal charging of electric-drive vehicles with renewable energy. Energy **36**(11), 6567–6576 (2011)
73. L. Zhu, F.R. Yu, B. Ning, Optimal charging control for electric vehicles in smart microgrids with renewable energy sources, in *IEEE Vehicular Technology Conference* IEEE (2012), pp. 1–5
74. T. Zhang, W. Chen, Z. Han, Z. Cao, Charging scheduling of electric vehicles with local renewable energy under uncertain electric vehicle arrival and grid power price. IEEE Trans. Veh. Technol. **63**(6), 2600–2612 (2014). future issue
75. C. Jin, X. Sheng, P. Ghosh, Energy efficient algorithms for electric vehicle charging with intermittent renewable energy sources, in *IEEE Power and Energy Society* General Meeting, IEEE (2013), pp. 1–5
76. C. Jin, X. Sheng, P. Ghosh, Optimized electric vehicle charging with intermittent renewable energy sources. IEEE J. Sel. Top. Signal Process. **8**(6), 1063–1072 (2014)

Chapter 3
Demand and Supply Management in Microgrids

3.1 Introduction

Microgrids are expected to be more robust and cost-effective than the traditional approach of centralized grids. However, a number of technical and regulatory issues have to be resolved before the microgrid can become a commonplace. One problem requiring due attention is the effective management of power supply and demand loads, which amounts to matching the power generation and consumption profiles [1, 2]. Specifically, the power generators or microsources employed in microgrids are usually renewable or non-conventional distributed energy resources. While incorporating such renewable resources shall bring great environmental benefits, it imposes new challenges as well: Different from that in the traditional power systems with conventional controllable electric generators, generation scheduling in microgrids with fluctuant, climate-dependent renewable energy sources has to cope with the non-trivial uncertainties.

The microgrids may adopt hierarchical or decentralized demand control schemes [3, 4]. The decentralized control schemes facilitate distributed control and management of large complex systems. However, such control requires significant experiments before implementation. Also, it may introduce new security challenges. Hierarchical control is performed by a master controller which is responsible for matching the generation and load. When the demand resources are controlled upon the occurrence of disturbance, the strategy is often known as direct load control [5, 6]. In a direct control program, based on an agreement between the central controller and customers, the controller can remotely control the operations of certain appliances in a household. This capability can be especially effective where there are electric devices allowing flexible usage time and/or energy storage, such as electric water heater (EWH) equipped with hot water storage tank and plug-in hybrid electric vehicles (PHEVs). The Kyotango microgrid project in Japan is an example of hierarchically controlled microgrid [7].

This chapter tackles the basic problem faced by the microgrid system central controller (MGCC), namely to achieve a good match between power demand and

© Springer Nature Singapore Pte Ltd. 2018

R. Wang et al., *Intelligent Microgrid Management and EV Control Under Uncertainties in Smart Grid*, https://doi.org/10.1007/978-981-10-4250-8_3

supply subject to uncertainties of renewable energy. On the power demand side, we envision a scenario with real-time communication between the controller and energy consumer premises. Specifically, in each time period, the operator controller receives consumer power demands with different power-level requirements, durations, and time elasticity levels. The MGCC needs to minimize the electricity generation cost by optimally scheduling the operation of each appliance subject to the requirements set by the users. Here, the generation fuel cost is modeled as a convex function of instantaneous total power consumption.

On the power supply side, MGCC has to focus on effectively managing power generation in order to match the user load and maintain system reliability. A novel uncertainty model is proposed to capture the fluctuant nature of renewable energy. Compared with previous robust optimization-based approaches which confine the renewable energy within a lower bound and an upper bound, the proposed model provides more statistical details in describing the underlying uncertainty. Specifically, an empirical distribution is extracted as a useful reference, which allows the actual distribution of renewable energy to vary around it. To the best of our knowledge, this is the first time that the distribution uncertainty model is adopted to depict the indeterminacy property of renewable energy generation. The load balance constraint is aptly approximated using the chance-constraint representation, which allows convenient tuning of the conservation level of the solution using a single parameter. A tractable robust optimization approach is developed for transforming the chance constraints into linear constraints and then solving the problem. It is shown that the proposed power demand and supply management scheme greatly reduces the energy cost for the microgrid system. Furthermore, some of the desirable properties of the proposed scheme are investigated, which sheds light on policy making for the future MGCC.

The remainder of this chapter is organized as follows. In Sect. 3.2, we show the mathematical depiction of the power demand and supply management problem and the uncertainty model of the renewable energy. Section 3.3 presents the robust approach for handling the load balance constraint. Simulation results and discussions are presented in Sect. 3.4. Finally, this chapter is concluded in Sect. 3.5.

3.2 Formulation of the Microgrid Demand and Supply Management Problem

In this section, a mathematical representation of the energy consumption and generation scheduling problem in an islanded microgrid with renewable energy is provided. An MGCC is responsible for scheduling the operations of the microgrid as well as performing optimization for minimizing the electricity generation cost for the microgrid system. The operations of the system and its mathematical depictions are introduced from the energy user side and energy generation side, respectively.

The uncertainty model for describing the randomness of renewable energy is then demonstrated.

3.2.1 Energy Demand Side

Consider a group of energy consumers participating in this energy consumption scheduling program. It is assumed that there are two-way communication infrastructures (e.g., a local area network (LAN)) between MGCC and energy consumers. Let $\mathcal{A}$ denote the set of flexible appliances belonging to these consumers [8], which may include PHEVs, dishwashers, cloth dryers, air conditioners. Time is divided into discrete time slots with equal length.[1] For each appliance a that is switched on, the active power consumed during one unit of time slot is x_a. An energy consumption scheduling vector $\mathbf{y_a}$ is also defined for each appliance a as follows:

$$\mathbf{y_a} = [y_a^1, ..., y_a^H] \tag{3.1}$$

where $H \geq 1$ is the scheduling horizon indicating the number of time slots ahead that are taken into account for decision-making in the energy consumption scheduling. For each coming time slot $h \in \mathcal{H} = [1, 2, ..., H]$, a binary variable $y_a^h = 0/1$ denotes the state of appliance a (on/off). Under such case, the actual energy consumption for appliance a at time slot h can be expressed as $x_a \cdot y_a^h$.

There is usually an upper limit on the total energy consumption in the microgrid in each time slot. Denoting this limit as E^{max}, we have:

$$\sum_{a \in \mathcal{A}} x_a \cdot y_a^h \leq E^{max}, \quad \forall h \in \mathcal{H}. \tag{3.2}$$

Next, assume that for each appliance $a \in \mathcal{A}$, the user indicates $\alpha_a, \beta_a \in \mathcal{H}$ as the beginning and end of a time interval in which the appliance a can be scheduled. Obviously, $\alpha_a < \beta_a$. For instance, the user may select $\alpha_a = 8$ PM and $\beta_a = 6$ AM (the next day) for his PHEV so that he could plug it in at night and get it fully charged before going to work the next day. Denote the minimum number of time slots needed for appliance a to finish its preset work as T_a. Given the predetermined parameters α_a, β_a, T_a, the appliance scheduling is subject to the following constraints:

$$\sum_{h=\alpha_a}^{\beta_a} y_a^h \geq T_a, \quad \forall a \in \mathcal{A}, \tag{3.3}$$

[1] The duration of one time slot is set as one hour in this chapter. Balancing periods of 5–30 minutes are also adopted in many countries [9]. Note that the length of one time slot is enough for solving the problem and the communication delay can be negligible as well.

and

$$y_a^h = 0, \quad \forall a \in \mathcal{A}, \quad \forall h \in \mathcal{H} \setminus [\alpha_a, \beta_a]. \tag{3.4}$$

Constraint (3.3) shows that the time length $\beta_a - \alpha_a$ needs to be large enough to allow finishing the normal operation of appliance a. In addition, the energy user can choose proper α_a, β_a, and T_a to indicate whether the operation of appliance a needs to be started immediately ($\beta_a - \alpha_a = T_a$) or can be deferred ($\beta_a - \alpha_a > T_a$).

To reveal the ramping-down and ramping-up limits on load levels of each time slot, we have:

$$\sum_{a \in \mathcal{A}} x_a \cdot y_a^h - \sum_{a \in \mathcal{A}} x_a \cdot y_a^{h+1} \leq r^D, \quad h \in [1, 2..., H-1], \tag{3.5}$$

$$\sum_{a \in \mathcal{A}} x_a \cdot y_a^{h+1} - \sum_{a \in \mathcal{A}} x_a \cdot y_a^h \leq r^U, \quad h \in [1, 2..., H-1]. \tag{3.6}$$

In this regard, it is assumed that each household participating in this energy consumption scheduling program is equipped with a smart meter, which is capable of detecting the electric power level of each appliance. The energy consumer announces to the MGCC his needs by selecting parameters α_a, β_a, and T_a for each appliance $a \in \mathcal{A}$.

The above constraints (3.2)–(3.6) describe common characteristics of household appliances. However, there exist some appliances of which the operation cannot be interrupted. Such kind of loads are called as uninterruptible loads. Discussions on how such loads may be handled are presented below.

Operation of Uninterruptible Loads: Some loads are interruptible, such as PHEV, which means that it is possible to charge the battery for some time, stop charging for some time, and then switch on the charging process again. Some other loads, however, are not interruptible, e.g., microwave oven. Appliances generating such loads, once started, have to be finished in one go. For each uninterruptible appliance $a \in \mathcal{A}'$, where $\mathcal{A}'$ represents the set of uninterruptible appliances, and each time slot h, let z_a^h denote an auxiliary binary variable such that $z_a^h \triangleq 1$ if appliance a starts operation at time slot h and $z_a^h \triangleq 0$ otherwise. We have

$$\sum_{h=\alpha_a}^{\beta_a - T_a + 1} z_a^h = 1 \tag{3.7}$$

and

$$z_a^h = 0, \quad \forall h \in \mathcal{H} \setminus [\alpha_a, \beta_a - T_a + 1]. \tag{3.8}$$

Then, we relate start time vector $\mathbf{z_a^h}$ with decision variable vector $\mathbf{y_a^h}$ as follows:

$$y_a^h \geq z_a^h, y_a^{h+1} \geq z_a^h, ..., y_a^{h+T_a-1} \geq z_a^h. \tag{3.9}$$

From (3.9), if $z_a^h = 1$, then $y_a^h = y_a^{h+1} = ... = y_a^{h+T_a-1} = 1$.

3.2.2 Energy Supply Side

We now turn to the energy supply side to consider the load balance constraint in the microgrid. The microgrid may be viewed as a graph consisting of three nodes as illustrated in Fig. 3.1. The first node represents the renewable energy generation sources such as wind turbines, solar panels, and fuel cells. At time slot h, denote the total energy generated in this node as ξ^h, where ξ^h is a random variable of which the probability density function may not be known. Node 2 in Fig. 3.1 represents the load connected through the transmission line to node 1 and node 3. The load at time h, denoted as l^h, is dependent on the energy consumption from the user side which, from the above analysis, can be expressed as:

$$l^h = \sum_{a \in \mathcal{A}} x_a \cdot y_a^h. \tag{3.10}$$

Finally, the third node includes a group of controllable electricity generators, which has a total amount of generation P_{cg}^h as commanded by MGCC. Controllable generators in microgrid typically include gas turbines, microturbines, reciprocating internal combustion engines with generators. These generators are powered by fossil fuels and can be controlled to compensate the mismatch between load and renewable power supply. A key requirement to the MGCC is to set the generation source power such that the supply could meet the demand. This statement can be mathematically described as

$$\xi^h + P_{cg}^h \geq l^h. \tag{3.11}$$

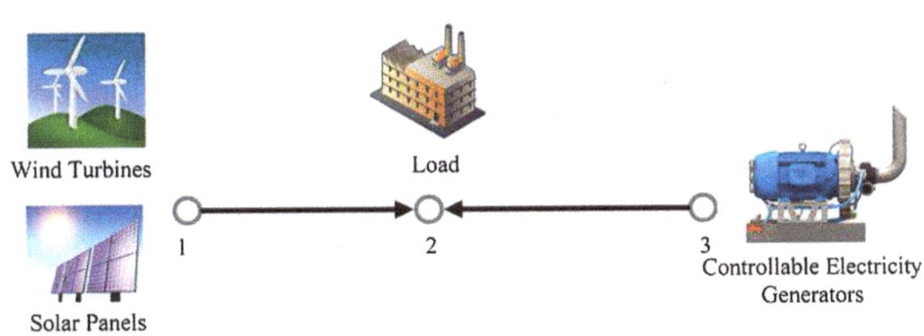

Fig. 3.1 The microgrid's graph

3.2.3 Problem Formulation

The objective function of MGCC can be defined in terms of minimizing the energy cost of the whole microgrid system. The optimal energy consumption scheduling problem therefore can be formulated as follows:

$$\begin{aligned} \min \quad & \sum_{h=1}^{H} \mathcal{C}_h(P_{cg}^h) \\ \text{s.t.} \quad & (3.2) \quad \text{to} \quad (3.11) \end{aligned} \tag{3.12}$$

where $\mathcal{C}_h(\cdot)$ is the cost function of electricity plant in the microgrid, which is assumed to be an increasing convex function. The convex property reflects the fact that each additional unit of power needed to serve the demands is provided at a higher cost. Example cases include the quadratic cost function [10, 11] and the piecewise linear cost function [1, 12]. Without loss of generality, we consider quadratic cost function $\mathcal{C}_h(P_{cg}^h) = a_h {P_{cg}^h}^2 + b_h P_{cg}^h + c_h$ throughout this chapter, where $a_h \geq 0$, $b_h \geq 0$, and $c_h \geq 0$ are known parameters for each time slot h. In practice, the coefficient of the quadratic term is usually small. Therefore, the quadratic cost function can be reduced to a linear cost function. As to the renewable energy cost, for typical renewable energy (e.g., solar and wind energy), capital cost dominates. The operation and maintenance costs are typically very low or even negligible [13, 14]. In this chapter, it is assumed that the renewable energy generators such as solar panels and wind turbines have already been installed, and the marginal cost of renewable energy can be neglected, leading to its omission in the objective function [15]. The main difficulty in solving problem (3.12) is the indeterminacy of renewable energy generation ξ^h existing in constraint (3.11). Note that optimizing over the space defined by (3.11) amounts to solving an optimization problem with potentially a large or even infinite number of constraints. Obviously, this realization of uncertainty is intractable. Next, a practical and flexible model will be developed to capture the uncertainty of ξ^h.

3.2.4 Probability Distribution Measure of Renewable Energy

It is generally difficult to characterize the renewable energy generation. In previous optimization approaches, operations on the random variable ξ^h is cumbersome and computationally intractable. Moreover, in practice, knowledge of the precise distribution of ξ^h may not be available. Solutions based on assumed distributions hence may not be justified. The variability of a random variable is usually measured using its variance or second moments which, however, may not provide sufficient details in describing the random variable. In this chapter, a reference distribution, rather than moment statistics, is extracted from historical data that will capture the distribution properties. Since renewable energy generation distribution is fluctuating over time and hard to be described in a closed-form expression, an empirical distribution may be adopted as a useful reference and allow the actual distribution to fluctuate around it. For example, it may be assumed that the renewable energy generation distribution $f_0(\xi^h)$ is shifting around a known Gaussian distribution (or other distribution) $g_h(\xi^h)$, which can be obtained based on long-term field measurements.

The discrepancy between $f_0(\xi^h)$ and its reference $g_h(\xi^h)$ can be described by a probabilistic distance measure, for example, the Kullback–Leibler (KL) divergence [16] which is a non-symmetric measure of the difference between two probability distributions. Name these two distributions as $f(\xi^h)$ and $g(\xi^h)$, respectively. Generally, one of the distributions, say, $f(\xi^h)$, represents the real distribution through precise modeling, while the reference $g(\xi^h)$ is a closed-form approximation based on the theoretic assumptions and simplifications. The definition of the KL divergence between two continuous distributions is given as follows:

$$D_{KL}(f(\xi^h), g(\xi^h)) = \int_{\xi^h \in S} [\ln f(\xi^h) - \ln g(\xi^h)] f(\xi^h) d\xi^h, \tag{3.13}$$

where S is the integral domain. When distributions $f(\xi^h)$ and $g(\xi^h)$ are close to each other, the distance measure is close to zero. Adopting the KL divergence, the distribution uncertainty set is defined as follows:

$$U_r(g(\xi^h), D_0) = \{f(\xi^h) \mid \mathbb{E}_f[\ln f(\xi^h) - \ln g(\xi^h)] \leq D_0\}, \tag{3.14}$$

where $D_0 \geq 0$ represents a distance limit and is obtained from empirical data or real-time measurement. It indicates energy generation's variation level. If the energy generation is very volatile, we have less confidence in the reference distribution and thus may set a larger distance limit.

Considering the renewable energy generation distribution $f_0(\xi^h)$ with reference distribution $g_h(\xi^h)$ and distance limit D_h, we have the following constraints for renewable energy generation distribution $f_0(\xi^h)$:

$$\mathbb{E}_{f_0}[\ln f_0(\xi^h) - \ln g_h(\xi^h)] \leq D_h, \tag{3.15}$$

$$\mathbb{E}_{f_0}[1] = 1. \tag{3.16}$$

Given (3.15) and (3.16), the load balance constraint (3.11) can be transformed to allow efficient solution of problem (3.12).

Remark Finding a proper reference distribution and obtaining an appropriate distance limit sometimes may be difficult, especially when still in lack of historical data. However, there are good reasons to expect that in most cases, situation could be improved quickly with continuous accumulation of historical records.

3.3 Optimization Algorithms

In this section, the optimization algorithms for solving the prime problem (3.12) are presented. Firstly, a robust approach for handling the load balance constraint is proposed. Then, the prime problem is decomposed into a sub-problem and a main problem to allow easier solution. Finally, the possible extensions of the proposed algorithm are briefly discussed.

3.3.1 Robust Approach for the Load Balance Constraint

As shown in (3.11), the load balance constraint is $\xi^h + P_{cg}^h \geq l^h$. In practice, a decision criterion is to set P_{cg}^h and l^h in such a way that people can be confident that the load balance constraint is achieved. To achieve that, we may introduce a small value ϵ to control the degree of conservatism and change the above expression into a chance constraint:

$$\mathbf{P}(\xi^h \leq l^h - P_{cg}^h) \leq \epsilon, \tag{3.17}$$

where ϵ is the fault tolerance limit of the power grid, representing the acceptable probability that the desirable power supply is not attained. Then, its robust expression can be obtained:

$$\max_{f_0(\xi^h) \in \boldsymbol{U}_r(g_h, D_h)} \mathbf{P}(\xi^h \leq l^h - P_{cg}^h) \leq \epsilon, \tag{3.18}$$

which is equivalent to:

$$\max_{f_0(\xi^h) \in \boldsymbol{U}_r(g_h, D_h)} \int_0^{l^h - P_{cg}^h} f_0(\xi^h) \mathrm{d}\xi^h \leq \epsilon. \tag{3.19}$$

Define $\delta^h = l^h - P_{cg}^h$ as the robust renewable energy usage (REU) decision, which equals the amount of energy dispatched to renewable energy plants at time slot h. In addition, an auxiliary function can be introduced as follows:

$$h(\xi^h, \delta^h) = \begin{cases} 1, & \xi^h \leq \delta^h; \\ 0, & \xi^h > \delta^h. \end{cases} \tag{3.20}$$

The left part of inequality (3.19) then can be formulated into an optimization problem:

$$\max_{f_0(\xi^h)} \int_0^{+\infty} h(\xi^h, \delta^h) \cdot f_0(\xi^h) \mathrm{d}\xi^h \tag{3.21}$$
$$\text{s.t.} \quad \mathbb{E}_{f_0}[\ln f_0(\xi^h) - \ln g_h(\xi^h)] \le D_h$$
$$\mathbb{E}_{f_0}[1] = 1$$

Define $Q_f^h(\delta^h) = \max_{f_0(\xi^h)\in U_r(g_h, D_h)} \int_0^{+\infty} h(\xi^h, \delta^h) \cdot f_0(\xi^h)\mathrm{d}\xi^h$ as the worst-case fault probability. We can then obtain a worst-case mapping $\mathcal{M}_{ws}^h$ which maps robust REU decision δ^h to $Q_f^h(\delta^h)$:

$$\mathcal{M}_{wc}^h : \quad \delta^h \longrightarrow Q_f^h(\delta^h). \tag{3.22}$$

3.3.2 Sub-Problem: Determine the Robust REU Decision Threshold

Since there exists a random variable ξ^h in the constraints, the energy generation and consumption scheduling problem (3.12) cannot be solved directly. As aforementioned, problem (3.12) can be decomposed into a sub-problem and a main problem. The goal of the sub-problem is to determine the robust REU decision threshold δ^{h*} so that the load balance constraint can be transformed into a solvable form.

Proposition 3.1 *Problem (3.21) is a convex optimization problem.*

Proof Rewrite (3.21) as follows:

$$\max_{f_0(\xi^h)} \int_0^{+\infty} h(\xi^h, \delta^h) \cdot f_0(\xi^h) \mathrm{d}\xi^h \tag{3.23}$$
$$\text{s.t.} \int_0^{+\infty} [\ln f_0(\xi^h) - \ln g_h(\xi^h)] f_0(\xi^h) \mathrm{d}\xi^h \le D_h \tag{3.24}$$
$$\int_0^{+\infty} f_0(\xi^h) \mathrm{d}\xi^h = 1. \tag{3.25}$$

It can be seen that the objective function (3.23) and equality constraint function (3.25) are affine with respect to $f_0(\xi^h)$. Next, it is shown that the inequality constraint function (3.24) is convex.

Lemma 3.1 *If* $f : \mathbf{R^n} \longrightarrow \mathbf{R}$ *is convex, then the perspective of* f*, which is denoted as a function* $g : \mathbf{R^{n+1}} \longrightarrow \mathbf{R}$ *that*

$$g(x, t) = tf(x/t), \tag{3.26}$$

with domain

$$\textbf{dom}\ g = \{(x, t)|x/t \in \textbf{dom}\ f,\ t > 0\} \tag{3.27}$$

preserves convexity.

That is to say, if f is a convex function, so is its perspective function g. Similarly, if f is concave, so is g. This can be proved in several ways, e.g., by direct verification of the defining inequality or using epigraphs and the perspective mapping on $\mathbf{R}^{\mathbf{n+1}}$. Readers can refer to [17] for more detailed discussions.

Consider the convex function $f(x) = -\ln x$ on $\mathbf{R}_{++}$. Its perspective is

$$g(x, t) = -t\ \ln(x/t) = t\ \ln(t/x) = t(\ln t - \ln x) \tag{3.28}$$

and it is convex on $\mathbf{R}^{\mathbf{2}}_{++}$. The function g is called the relative entropy of t and x. Then, we have that the KL divergence $\int_{x\in S}[\ln f(x) - \ln g(x)]f(x)\mathrm{d}x$ between distribution $f(x)$ and $g(x)$ is convex in $f(x)$ (and $g(x)$ as well). In this case, it is claimed that the inequality constraint (3.24) is convex in distribution $f_0(\xi^h)$.

Through Slater's condition, strong duality holds for problem (3.23)–(3.25). Adopting the Lagrangian method, the worst-case fault probability $Q^h_f(\delta^h)$ can be obtained as follows:

$$Q^h_f(\delta^h) = \min_{\tau,\eta}\max_{f_0(\xi^h)} \mathbb{E}_{f_0}\left[h(\xi^h, \delta^h) - \eta - \tau\ln\frac{f_0(\xi^h)}{g_h(\xi^h)}\right] + \tau D_h + \eta$$

where $\tau \geq 0$ and η are Lagrangian multipliers associated with constraints (3.24)–(3.25), respectively. Let

$$\mathcal{P}(\delta^h, f_0, \tau, \eta) = \mathbb{E}_{f_0}\left[h(\xi^h, \delta^h) - \eta - \tau\ln\frac{f_0(\xi^h)}{g_h(\xi^h)}\right], \tag{3.29}$$

the derivative of $\mathcal{P}(\delta^h, f_0, \tau, \eta)$ with respect to f_0 is as follows:

$$\begin{aligned}\frac{\partial\mathcal{P}}{\partial f_0} &= \lim_{t\to 0}\frac{1}{t}\Big[\mathcal{P}\big(f_0(\xi^h) + t\cdot g_0(\xi^h)\big) - \mathcal{P}\big(f_0(\xi^h)\big)\Big] \\ &= \int_0^{+\infty}\left(h(\xi^h, \delta^h) - \tau\ln\frac{f_0(\xi^h)}{g_h(\xi^h)} - \eta - \tau\right) g_0(\xi^h)\mathrm{d}\xi^h.\end{aligned} \tag{3.30}$$

Using the Karush–Kuhn–Tucker (KKT) optimality conditions, we have

$$h(\xi^h, \delta^h) - \tau\ln\frac{f_0(\xi^h)}{g_h(\xi^h)} - \eta - \tau = 0 \tag{3.31}$$

$$\int_0^{+\infty} f_0(\xi^h)\mathrm{d}\xi^h = 1 \tag{3.32}$$

$$\mathbb{E}\left[\ln \frac{f_0(\xi^h)}{g_h(\xi^h)}\right] - D_h \leq 0 \tag{3.33}$$

$$\tau \cdot \left(D_h - \mathbb{E}\left[\ln \frac{f_0(\xi^h)}{g_h(\xi^h)}\right]\right) = 0 \tag{3.34}$$

$$\tau \geq 0 \tag{3.35}$$

From (3.31), the optimal distribution function can be expressed as follows:

$$f_0^*(\xi^h) = g_h(\xi^h) \exp\left(\frac{h(\xi^h, \delta^h) - \eta}{\tau} - 1\right). \tag{3.36}$$

The dual variables (τ, η) in (3.36) should be chosen properly such that conditions (3.32)–(3.35) are satisfied. Specifically, the following results can be obtained:

Proposition 3.2 *The choice of* (τ, η) *is a solution of the following nonlinear equations:*

$$H_1(\tau, \eta) = R(\delta^h)e^{-\eta/\tau} + S(\delta^h)e^{(1-\eta)/\tau} - 1 = 0 \tag{3.37}$$

$$H_2(\tau, \eta) = S(\delta^h)e^{(1-\eta)/\tau} - \eta - \tau(1 + D_h) = 0, \tag{3.38}$$

where $S(\delta^h) = (1 - G_h(\delta^h)) \exp(-1)$, $R(\delta^h) = G_h(\delta^h) \exp(-1)$, *and* $G_h(\delta^h) = \int_{\xi^h \geq \delta^h} g_h(\xi^h) d\xi^h$ *denotes the complementary cumulative distribution function of reference distribution* $g_h(\xi^h)$.

Proof By substituting the optimal distribution $f_0^*(\xi^h)$ back into (3.32) and $f_0^*(\xi^h)$, (3.31) into (3.34), we have

$$\int_0^{+\infty} g_h(\xi^h) \exp\left(\frac{h(\xi^h, \delta^h) - \eta}{\tau} - 1\right) \mathrm{d}\xi^h = 1 \tag{3.39}$$

$$\begin{aligned}&\int_0^{+\infty} \left(h(\xi^h, \delta^h) - \eta - \tau\right) g_h(\xi^h) \\ &\cdot \exp\left(\frac{h(\xi^h, \delta^h) - \eta}{\tau} - 1\right) \mathrm{d}\xi^h - D_h \cdot \tau = 0,\end{aligned} \tag{3.40}$$

which are equivalent to:

$$\begin{aligned}&\exp\left(-1 - \frac{\eta}{\tau}\right) \cdot \int_{\delta^h}^{+\infty} g(\xi^h) \mathrm{d}\xi^h \\ &+ \exp\left(-1 + \frac{1-\eta}{\tau}\right) \cdot \int_0^{\delta^h} g(\xi^h) \mathrm{d}\xi^h - 1 = 0\end{aligned} \tag{3.41}$$

$$(1-\eta-\tau)\exp\left(-1+\frac{1-\eta}{\tau}\right)\cdot\int_0^{\delta^h} g(\xi^h)\mathrm{d}\xi^h \\ +(-\eta-\tau)\exp\left(-1-\frac{\eta}{\tau}\right)\int_{\xi^h}^{\infty} g(\xi^h)\mathrm{d}\xi^h - \tau D_h = 0. \tag{3.42}$$

Equation (3.37) can be easily obtained from (3.41) by introducing $S(\delta^h)$ and $R(\delta^h)$. Through (3.41), (3.42) can be transformed into:

$$(1-\eta-\tau)\exp\left(-1+\frac{1-\eta}{\tau}\right)\cdot\int_0^{\delta^h} g(\xi^h)\mathrm{d}\xi^h + \\ (-\eta-\tau)\left[1-\exp\left(-1+\frac{1-\eta}{\tau}\right)\cdot\int_0^{\delta^h} g(\xi^h)\mathrm{d}\xi^h\right] - \tau D_h = 0.$$

Then

$$\exp\left(-1+\frac{1-\eta}{\tau}\right)\cdot\int_0^{\delta^h} g(\xi^h)\mathrm{d}\xi^h - \eta - \tau - \tau D_h = 0, \tag{3.43}$$

which is equivalent to (3.38). Hence, Proposition 3.2 is proved.

It is, however, still rather difficult to obtain an explicit solution from (3.37) and (3.38). Hence, we propose the Newton iterations as detailed in Algorithm 3.1.

Once the solutions for (3.37)–(3.38) in Proposition 3.2 are determined, through (3.31)–(3.34), the worst-case fault probability can be obtained as follows:

$$Q_f^h(\delta^h) = \mathbb{E}_{f_0^*}[h(\xi^h, \delta^h)] = (1+D_h)\tau + \eta \tag{3.44}$$

Our next step is then to find the robust REU decision threshold δ^{h*} such that $Q_f^h(\delta^{h*}) = \epsilon$, which involves the calculation of inverse function of $Q_f^h(\delta^h)$, and it is not directly possible from (3.44). The following property of function $Q_f^h(\delta^h)$, however, may help us design such a search method.

Proposition 3.3 *The worst-case fault probability $Q_f^h(\delta^h)$ is non-decreasing with respect to the REU decision δ^h.*

The conclusion in Proposition 3.3 is straightforward since

$$dQ_f^h(\delta^h)/d\delta^h = d\mathbb{E}_{f_0^*}[h(\xi^h, \delta^h)]/d\delta^h = f_0^*(\delta^h) \geq 0.$$

Though direct solution is not available, the monotonicity of $Q_f^h(\delta)$ enlightens a bisection method to search for the solution for $Q_f^h(\delta^h) = \epsilon$. The main idea is to perform the search within an interval of $[0, \rho]$, where ρ is an empirical constant such that $Q_f^h(\rho) > \epsilon$.

Details of the algorithm for searching the robust REU decision threshold are presented in Algorithm 3.1. Note that, from the 3rd to the 11th lines of the algorithm, Newton iteration is adopted to solve the equation in Proposition 3.2 and obtain the worst-case probability with fixed robust REU decision. Then, the worst-case probability at $\delta^h{}_-$ and δ^{h^-} is compared with the fault tolerant limit ϵ, respectively. The comparison results help shrink the search region as shown in lines 12–14.

Algorithm 3.1 Search for robust REU decision threshold δ^{h^*}

Input: Reference distribution $g_h(\xi^h)$;
Distance limit D_h;
Search radius ρ;
Load balance fault tolerant limit ϵ;
Tolerance ε.
Output: Robust REU decision threshold such that $Q^h_f(\delta^{h^*}) = \epsilon$;
1: **Begin**
2: **initialize** $\delta^h{}_- = 0$, $\delta^{h^-} = \rho$, and set $\mathbf{H}(\tau, \eta) = [H_1(\tau, \eta), H_2(\tau, \eta)]^T$
3: **while** $|\delta^h{}_- - \delta^{h^-}| > \varepsilon$
4: **set** $\bar{\delta^h} = \frac{\delta^h{}_- + \delta^{h^-}}{2}$ and initiate the time iteration $k = 1$
5: **while** $\mathbf{H}(\tau, \eta) > \varepsilon$
6: **evaluate** $\mathbf{H}(\tau, \eta)$ and Jacobian matrix $\mathbf{J}(\tau, \eta)$
7: **solve** $\mathbf{J}(\tau, \eta)\Delta \mathbf{x_k} = -\mathbf{H}(\tau, \eta)$
8: **update** $\tau_{k+1} = [\tau_k + \Delta\tau_k]^+$, $\eta_{k+1} = \eta_k + \Delta\eta_k$
9: **update** $Q^h_f(\bar{\delta^h}) = (1 + D_h)\tau_{k+1} + \eta_{k+1}$
10: **set** $k = k + 1$
11: **end while**
12: **if** $\left(Q^h_f(\bar{\delta^h}) - \epsilon\right)\left(Q^h_f(\delta^{h^-}) - \epsilon\right) < 0$
13: **then** set $\delta^h{}_- = \bar{\delta^h}$ **else** set $\delta^{h^-} = \bar{\delta^h}$ **end if**
14: **if** $|Q^h_f(\bar{\delta^h}) - \epsilon| < \varepsilon$ break **end if**
15: **end while**
16: **set** $\delta^{h^*} = \bar{\delta^h}$
17: **End**

3.3.3 Main Problem: Determine the Optimal Energy Consumption and Generation Scheduling

Once the robust REU decision threshold δ^{h^*} for the robust load balance constraint (3.19) is obtained, the energy generation and consumption management problem can be reformulated. Specifically, the following optimization problem can be tackled rather than the original Eq. (3.12)

$$\min \sum_{h=1}^{H} C_h(P_{cg}^h)$$
$$\text{s.t.} \sum_{a \in \mathcal{A}} x_a \cdot y_a^h - P_{cg}^h = \delta^{h*}, \quad \forall h \in \mathcal{H},$$
$$\text{and} \quad (3.2) \quad \text{to} \quad (3.10), \tag{3.45}$$

where the optimization variables include the controllable energy generation variable P_{cg}^h for all time slots $h \in \mathcal{H}$ and the energy consumption scheduling vector $\mathbf{y_a}$ for all appliances $a \in \mathcal{A}$. The objective function aims at minimizing the overall energy cost in microgrid over the whole time horizon.

It can be seen that all the constraints of (3.45) are linear and the objective function is quadratic. This problem is a mixed-integer quadratic programming problem. Algorithms that can be adopted to tackle this kind of problem include the cutting plane method and the branch and bound method. This problem can also be effectively solved by some commercial optimization software including CPLEX, Mosek, FortMP and Gurobi.

3.3.4 Extensions of the Proposed Algorithm: A Brief Discussion

With trivial or, sometimes, non-trivial extensions, the proposed algorithm may be applied to solve some other power demand and supply problems in microgrids. A few possible extensions are briefly discussed as follows.

- The scenario that has been considered in this chapter assumes that the end users control their power consumption in accordance with the guideline that MGCC suggests. Under such case, the uncertainties from the end user side are expected to be limited and can be handled by the system, since the control sequences obtained by the proposed algorithm are of reasonably good robustness. For the cases where uncertainties from the user side exceed the range of tolerance (e.g., many end users do not follow the guideline for whatever reasons), the proposed uncertainty model can be extended to include the uncertainty from the end user side by properly integrating user-side and supply-side uncertainties. Once the integrated reference distribution of the combined uncertainties is obtained, the proposed algorithm can be used with virtually no changes. Detailed discussions on such cases, however, are out of the scope of this book. For the even worse case where unexpected real-time changes go beyond what have been modeled, a few classic approaches may be adopted, e.g., by increasing the power output of energy generators, turning on stand-by fast response generators, importing electricity from the power grid, or shedding load if necessary, etc.
- The power demand and supply management framework discussed so far is an offline approach suitable for planning the energy consumption and generation

ahead of time. When real-time adjustment is of a big concern yet response time limit is not too rigid, the proposed algorithm can be easily extended to handle such cases. One option is to adopt the model predictive control (MPC) approach (also known as "receding horizon approach") [18, 19], of which the basic idea is to calculate the optimal control sequences yet implement only the first step of them. In other words, the power demand and supply management problem is solved at time $h = \tau$ ($\tau \in \mathcal{H}$ denotes the current time index) for the remaining horizon $[\tau, \tau+1, \ldots, H]$, yet only the solution for the current time slot τ is implemented. In the next time slot, MGCC shall update the system information (e.g., the requirements of end users and robust REU decision thresholds) and redo the calculations. For the cases where response time has to be very short, however, different algorithms with lower complexities and faster speed probably have to be developed, in order to support real-time operation and small-step scheduling more efficiently. The designing of such algorithms shall be considered in our future studies. Note that, for such cases, enhancing the system fault tolerance against the noises of real-time data shall also be considered.

- The problem formulation can also be easily extended to handle the case where microgrids import electricity from outside power grid. Specifically, assuming that the microgrid imports P_E^h units of electricity from the power grid, the problem formulation (3.45) can be modified as:

$$\begin{aligned} \min \ & \sum_{h=1}^{H} a_h \cdot {P_{cg}^h}^2 + b_h \cdot P_{cg}^h + c_h + d_h \cdot P_E^h \\ \text{s.t.} \ & \sum_{a \in \mathcal{A}} x_a \cdot y_a^h - P_{cg}^h = \delta^{h*} + P_E^h, \quad \forall h \in \mathcal{H}, \\ & \text{and} \quad (3.2) \quad \text{to} \quad (3.10), \end{aligned} \tag{3.46}$$

where d_h is the electricity price of power grid at time slot h. The problem is essentially still a mixed-integer quadratic programming problem which can be solved by using the same algorithm, whereas if electricity price from the power grid d_h is time varying with non-trivial uncertainty, the problem will become more complicated. A feasible option is to develop a worst-case robust optimization approach for the problem. Readers may refer to [20–22] for more exhaustive descriptions on dealing with bounded uncertainties in the coefficients of objective function. Detailed discussions are not too difficult yet rather lengthy, and therefore have to be left to a separate report.

3.4 Simulation Results and Discussions

In this section, simulation results are presented for assessing the performance of the proposed power demand and supply management scheme and evaluating the effects of different system parameters. Here, an assumption is made on top of Refs. [23, 24], where Gaussian random process has been adopted to describe the renewable energy generation. Specifically, it is assumed that the reference distribution is a Gaussian distribution $g_h(\xi^h)$ with mean $\overline{m}_h$ and standard deviation σ_h. In addition, the parameters of the cost function in (3.45) for each time slot are set as $a_h > 0$, $b_h = 0$, and $c_h = 0$.

3.4.1 The Impacts of Distribution Uncertainty Set

We first set the fault tolerant limit $\epsilon = 10^{-3}$ and investigate the relations between robust REU decision threshold δ^{h*} and distance limit D_h for different values of $\overline{m}_h$ and σ_h. The results are plotted in Figs. 3.2 and 3.3. It is shown that the robust REU decision threshold decreases with the increase in the distance limit. This observation is intuitive since a larger distance limit defines a larger distribution set which allows the renewable energy output to fluctuate more intensively. Given the required fault tolerant limit, REU decision threshold has to be set at a lower value so as to have less reliance on renewable energy with stronger uncertainties and guarantee the system reliability. Note that when $D_h = 0$, the renewable energy follows exactly the reference distribution $g_h(\xi^h)$. In this special case, renewable energy generation is a random variable with determinate distribution $g_h(\xi^h)$. While $D_h > 0$, our reference model considers a more general case which allows discrepancy between actual

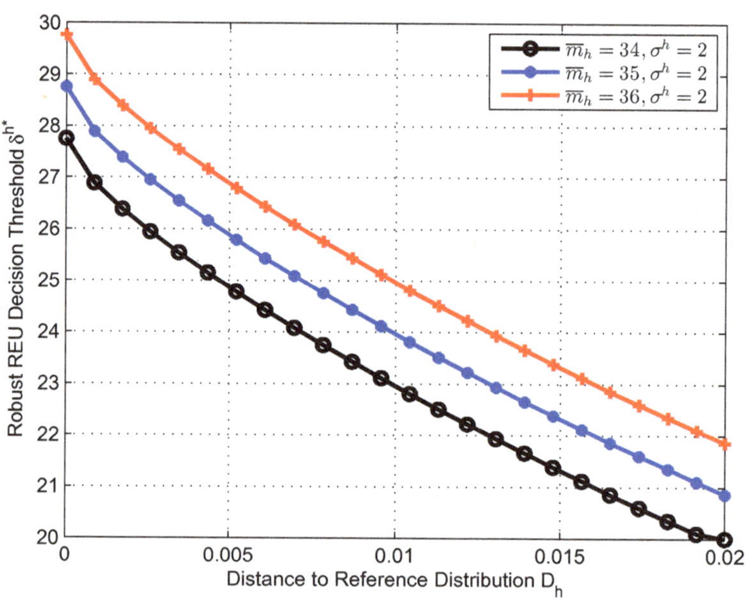

Fig. 3.2 Robust REU decision threshold δ^{h*} with distance limit D_h for different $\overline{m}_h$

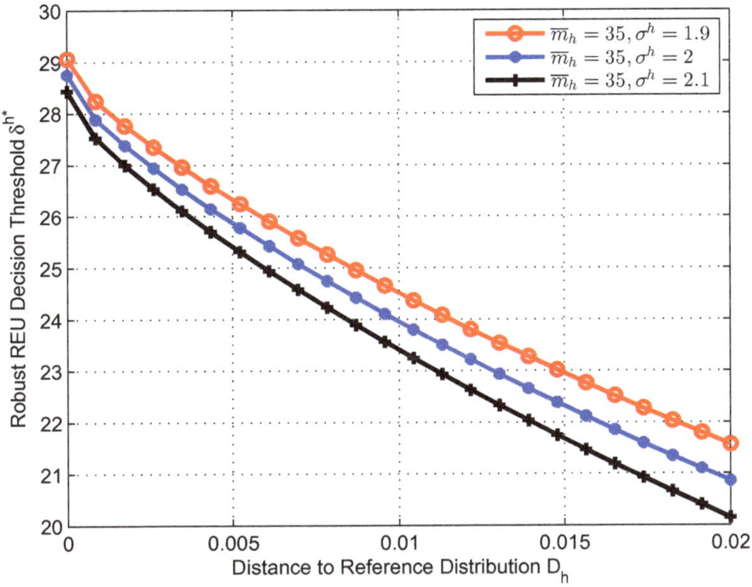

Fig. 3.3 Robust REU decision threshold δ^{h^*} with distance limit D_h for different σ^h

distribution and its reference. The discrepancy however is limited and confined by a probabilistic distance measure. Simply put, the reference model allows the actual renewable energy generation to follow a different distribution function from the reference distribution, but not be too disparate based on historical data or empirical knowledge.

From Figs. 3.2 and 3.3, the following statement can also be proposed: When the reference distribution is Gaussian, the robust REU decision threshold δ^{h^*} linearly increases with the mean of reference distribution $\overline{m}_h$ and linearly decreases with the standard deviation of reference distribution σ_h. This statement can be explained analytically as follows. $G_h(\delta^h)$ in (3.37)–(3.38) is first transformed into:

$$G_h(\delta^h) = \int_{\xi^h \geq \delta^h} g_h(\xi^h) \mathrm{d}\xi^h = \int_{\frac{\delta^h - \overline{m}_h}{\sigma_h}}^{+\infty} n(x) \mathrm{d}x, \quad (3.47)$$

where $n(x)$ is the probability density function of the standard Gaussian distribution. Since fault tolerant limit ϵ is of a relatively small value, we have that δ^{h^*} is less than $\overline{m}_h$. As $\overline{m}_h$ and σ_h vary, in order to preserve the same worst-case probability $Q_f^h(\delta^{h^*})$, the solutions (η, τ) of the Eqs. (3.37)–(3.38) need to remain unchanged, indicating that $S(\delta^{h^*})$, $R(\delta^{h^*})$ and $G(\delta^{h^*})$ also need to be constants. In this regard:

$$\frac{\delta^{h^*} - \overline{m}_h}{\sigma_h} = C \quad \Longrightarrow \quad \delta^{h^*} = C\sigma_h + \overline{m}_h,$$

where C is a negative constant. Thus, δ^{h^*} linearly increases with $\overline{m}_h$ and linearly decreases with σ_h.

3.4.2 Effects of Fault Tolerant Limit ϵ

We set $\overline{m}_h = 36$ and $\sigma_h = 2$ and investigate how the robust REU decision threshold varies when fault tolerance limit increases. Figure 3.4 plots the mapping from fault tolerant limit ϵ to robust REU decision threshold δ^{h^*} under different values of the distance limit D_h. The figure indicates that a larger fault tolerant limit permits a higher reliance on renewable energy (a larger robust REU decision threshold), which is straightforward to understand. Also note that the worst-case fault probability is an increasing function of the REU decision threshold. Thus, it is justified to adopt the bisection method as presented in Algorithm 3.1 to search for the REU decision threshold which satisfies the fault tolerant limit requirement. Note that in this figure, the red triangle line is the special case where renewable energy follows reference distribution exactly. It is also observed that robust REU threshold δ^{h^*} is more sensitive to ϵ when D_h increases.

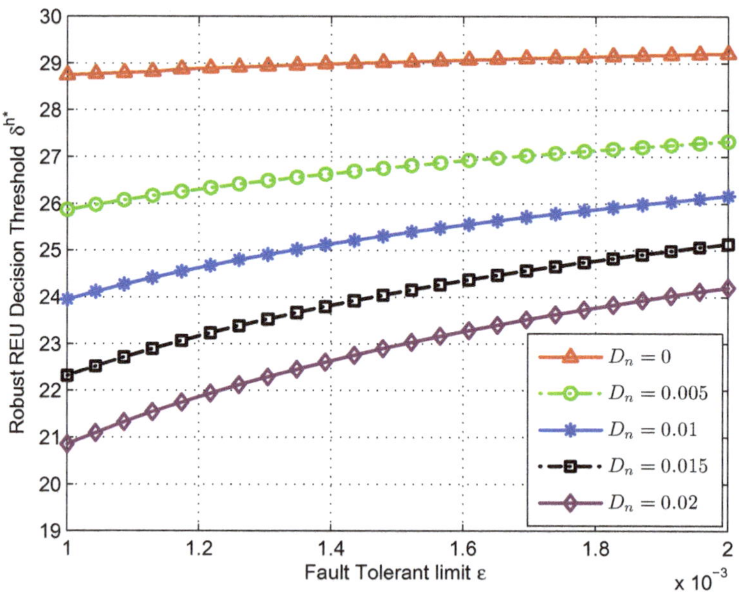

Fig. 3.4 Robust REU decision threshold δ^{h^*} with fault tolerant limit ϵ for different D_h

3.4.3 The Impacts of Uninterruptible Loads

For the experiment studied in this part, we set the power consumption scheduling horizon $|\mathcal{H}| = 12$ h. That is, the MGCC solves optimization problem (3.45) to decide on the operations of each appliance for the next 12 hours. In this chapter, 30 household appliances including electric cookers (EC), air conditioners (AC), electric water heaters (EWH), cloth dryers (CD), dishwashers (DW), and plug-in hybrid electric vehicles (PHEVs) are considered to study the optimal power consumption scheduling with a mixed-integer quadratic programming approach. The detailed operation data used in this chapter is modified based on the information from [25–29] and is shown in Table 3.1. Note that a user elasticity index is also introduced as follows:

$$\gamma_a = \frac{T_a}{\beta_a - \alpha_a + 1}$$

to describe the scheduling flexibility of appliance a. Obviously, $\gamma_a \in (0, 1]$, and a larger γ_a implies a more inflexible arrangement property. The operation window $[\alpha_a, \beta_a]$ of each appliance is chosen according to the preferences of different users. In this chapter, the values of α_a and β_a are not enumerated one by one due to limited space; instead, the ranges of γ_a for each kind of appliance are listed, which is presented in the last column of Table 3.1.

The mean $\overline{m}_h$ and standard deviation σ_h of the reference distribution, together with the distance limits for the next 12 time slots, are given in Table 3.2. Based on these data and adopting Algorithm 3.1, the robust REU threshold δ^{h*}, representing the amount of energy dispatched to renewable energy plants for each time slot, is obtained. The results are demonstrated in the last column of Table 3.2 and are used to solve the main problem (3.45). Our experiments utilize MOSEK optimization toolbox 6.0 on an Intel-P4 2.4-GHz personal computer. To investigate the impacts of uninterruptible loads, the following cases are studied:

- Case 1: Only electric cookers are classified into the uninterruptible appliance set $\mathcal{A}'$, i.e., $\mathcal{A}' = \{\text{EC}\}$.
- Case 2: On top of Case 1, air conditioners are added to the uninterruptible appliance set $\mathcal{A}'$, i.e., $\mathcal{A}' = \{\text{EC, AC}\}$.

Table 3.1 Operation data for appliances in the microgrid

Type of Appliance	Power Level (KW)	T_a	γ_a
EC	2	1	0.3–0.4
AC	3.5	10	0.9–1.0
EWH	4.5	3	0.5–0.7
CD	5	3	0.3–0.4
DW	0.85	2	0.3–0.4
PHEV	7.3	7	0.6–0.7

Table 3.2 Parameters of distribution uncertainty set and corresponding robust REU decision threshold

Time Slot	$\overline{m}_h$	σ_h	D_h	δ^{h*}
1	14.678	0.9571	0.0162	8.419
2	14.757	0.4853	0.0181	11.453
3	14.743	0.8002	0.0025	11.531
4	14.392	0.1418	0.0182	13.423
5	14.655	0.4217	0.0126	12.137
6	14.171	0.9157	0.0019	10.630
7	14.706	0.7922	0.0055	10.995
8	14.031	0.9594	0.0109	8.577
9	14.276	0.6557	0.0191	9.716
10	14.046	0.0357	0.0192	13.797
11	14.097	0.8491	0.0031	10.565
12	14.823	0.9339	0.0194	8.294

- Case 3: On top of Case 2, electric water heaters are added to the uninterruptible appliance set $\mathcal{A}'$, i.e., $\mathcal{A}' = \{\text{EC, AC, EWH}\}$.
- Case 4: On top of Case 3, cloth dryers are added to the uninterruptible appliance set $\mathcal{A}'$, i.e., $\mathcal{A}' = \{\text{EC, AC, EWH, CD}\}$.
- Case 5: On top of Case 4, dishwashers are added to the uninterruptible appliance set $\mathcal{A}'$, i.e., $\mathcal{A}' = \{\text{EC, AC, EWH, CD, DW}\}$.
- Case 6: On top of Case 5, PHEVs are added to the uninterruptible appliance set $\mathcal{A}'$, i.e., $\mathcal{A}' = \{\text{EC, AC, EWH, CD, DW, PHEV}\}$.

Figure 3.5 demonstrates the energy cost for each case. Obviously, the energy cost goes up when the uninterruptible appliance set $\mathcal{A}'$ is scaled up. We compare the costs of adjacent cases, and the difference between these costs is called cost gap. The largest cost gap is shown between Case 5 and Case 6 due to PHEVs' high electric power consumption ($P = 7.3$ KW) and relatively considerable scheduling elasticity ($\gamma_a = 0.6$–0.7).

3.4.4 The Price of User Elasticity

In this section, the effects of user elasticity on the energy cost of the microgrid system are explored. First, it is assumed that all the appliances are uninterruptible. Since the minimum running time of each appliance T_a is fixed, we extend or shrink the operation window $[\alpha_a, \beta_a]$ to relax or tighten the user elasticity. Note that, at one time, the operation window $[\alpha_a, \beta_a]$ of each appliance a will expand or shrink one unit of time slot from both sides; i.e., the operation window will scale up or down two time slots. If one side of the operation window cannot be extended due to the finite

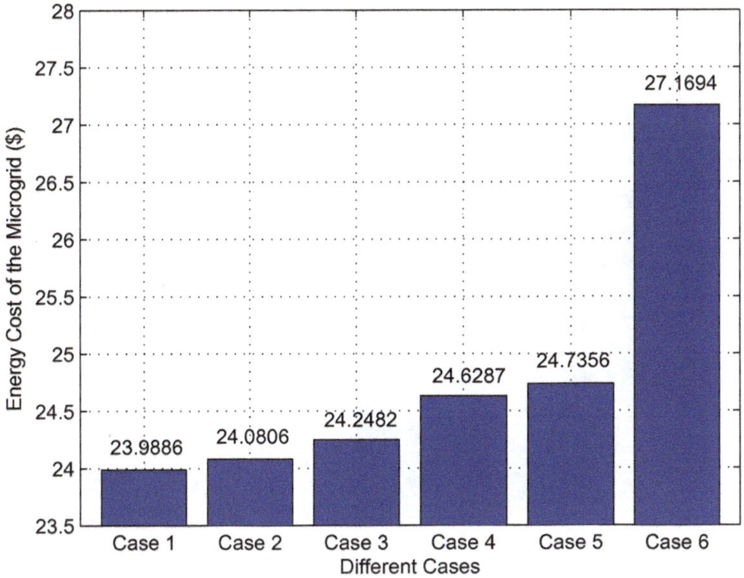

Fig. 3.5 The impacts of uninterruptible loads on the energy cost of the microgrid

length of time horizon, the operation window will only scale up on one side until it covers the whole time horizon. The operation window of each appliance is kept on extending or shrinking until all the operation windows cannot be changed. Then, how energy cost changes when operation windows vary is demonstrated. Note that the cases when operation windows are shrunk 6, 4, 2 time slots and extended 0, 2, 4 time slots are selected, respectively. The results are shown in Fig. 3.6. In this figure, it is observed that when user elasticities are tightened, energy cost increases rapidly. Such phenomenon can be interpreted that the user elasticity can make a significant impact on the energy cost of the microgrid system. Compared with the effects of interruptibility property, user elasticity has stronger influences on the expenditure of the whole system. This result may give MGCC an inspiration that it is worthy to provide more rewards to customers who agree to have more time flexibility than those who allow interruptions to some appliances. Moreover, it is also shown that when the operation windows are shrunk by 6 time slots, nearly all the appliances' elasticity reaches 1. This approximates the case when all the appliances operate at their desired time with no flexibility. Compared with this benchmark case, it can be observed that the proposed power consumption management scheme can reduce the energy cost significantly.

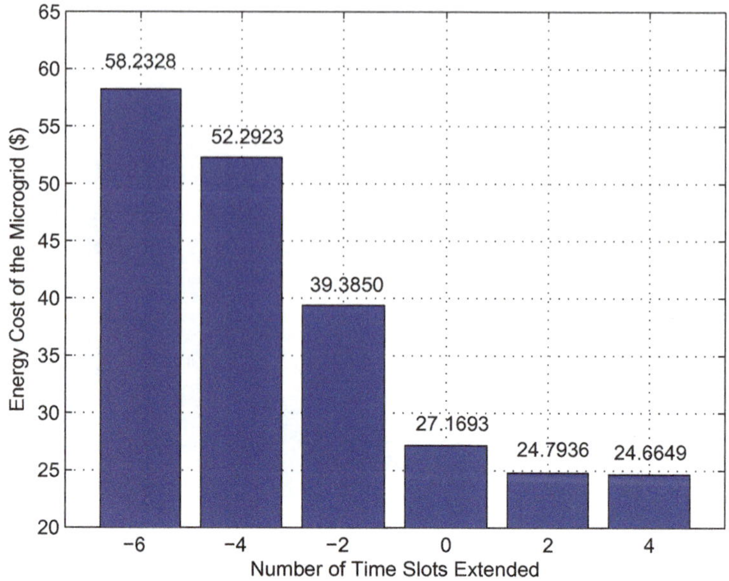

Fig. 3.6 The impact of user elasticity on the energy cost of the microgrid

3.5 Conclusion

In this chapter, a fundamental problem of using a microgrid system central controller to optimally schedule the demand and supply profiles so as to minimize the fuel consumption costs during the whole time horizon is studied. We focus on a scenario where the end-users control their power consumptions in accordance with the guideline that MGCC suggests. To tackle the randomness of renewable energy, a reference distribution is introduced and then a distribution uncertainty set is defined to confine the uncertainty. Such a novel model allows convenient handling of fluctuating renewable generation as long as the renewable energy generation profile is not too drastically different from the past observation or empirical knowledge. An optimization formulation of the problem is proposed, and a two-stage algorithm approach is developed, to firstly transform and then solve the problem. Numerical results indicate that the proposed energy consumption management scheme can significantly cut down energy expenses. Effects of a few factors, including the reference distribution, the fault tolerant limit, the types and amount of uninterruptible loads, and the user elasticity are carefully evaluated. Such evaluations, as we believe, help provide some useful insights for the developments of more effective payback policies for the future MGCC.

References

1. I. Koutsopoulos, L. Tassiulas, Optimal control policies for power demand scheduling in the smart grid. IEEE J. Select. Areas Commun. **30**(6), 1049–1060 (2012)
2. M.E. Khodayar, M. Barati, M. Shahidehpour, Integration of high reliability distribution system in microgrid operation. IEEE Trans. Smart Grid **3**(4), 1997–2006 (2012)
3. A.G. Tsikalakis, N.D. Hatziargyriou, Centralized control for optimizing microgrids operation. IEEE Trans. Energy Convers. **23**(1), 241–248 (2008)
4. J. Guerrero, P. Loh, M. Chandorkar, T. Lee, Advanced control architectures for intelligent microgrids, part i: decentralized and hierarchical control: advanced control architectures for intelligent microgrids. IEEE Trans. Ind. Electron. **60**(4), 1254–1262 (2013)
5. N. Ruiz, I. Cobelo, J. Oyarzabal, A direct load control model for virtual power plant management. IEEE Trans. Power Syst. **24**(2), 959–966 (2009)
6. A. Gomes, C. Antunes, A. Martins, A multiple objective approach to direct load control using an interactive evolutionary algorithm. IEEE Trans. Power Syst. **22**(3), 1004–1011 (2007)
7. M. Barnes, J. Kondoh, H. Asano, J. Oyarzabal, G. Ventakaramanan, R. Lasseter, N. Hatziargyriou, T. Green, Real-world microgrids-an overview, in *IEEE International Conference on System of Systems Engineering* (IEEE, 2007) pp. 1–8
8. L. Zheng, L. Cai, A distributed demand response control strategy using lyapunov optimization. IEEE Trans. Smart Grid **5**(4), 2075–2083 (2014)
9. C. Harris, *Electricity markets: pricing, structures and economics*. (Wiley, 2006)
10. A. Mohsenian-Rad, V. Wong, J. Jatskevich, R. Schober, A. Leon-Garcia, Autonomous demand-side management based on game-theoretic energy consumption scheduling for the future smart grid. IEEE Trans. Smart Grid **1**(3), 320–331 (2010)
11. P. Samadi, H. Mohsenian-Rad, R. Schober, V. Wong, Advanced demand side management for the future smart grid using mechanism design. IEEE Trans. Smart Grid **3**(3), 1170–1180 (2012)
12. C. Joe-Wong, S. Sen, S. Ha, M. Chiang, Optimized day-ahead pricing for smart grids with device-specific scheduling flexibility. IEEE J. Sel. Areas Commun. **30**(6), 1075–1085 (2012)
13. Renewable power generation costs in 2012: An overview (2012)
14. Handbook for solar photovoltaic (pv) systems (2011)
15. Y. Guo, M. Pan, Y. Fang, Optimal power management of residential customers in the smart grid. IEEE Trans. Parallel Distrib. Syst. **23**(9), 1593–1606 (2012)
16. R.M. Gray, *Entropy and information theory*. (Springer Science and Business Media, 2011)
17. S. Boyd, L. Vandenberghe, *Convex optimization*. (Cambridge university press, Cambridge 2004)
18. E.F. Camacho, C.B. Alba, *Model Predictive Control* (Springer, London, 2004)
19. F. Allgöwer, A. Zheng, *Nonlinear model predictive control*, vol. 26. (Birkhäuser Basel, 2000)
20. D. Bertsimas, M. Sim, Robust discrete optimization and network flows. Math. Program. **98**(1), 49–71 (2003)
21. D. Bertsimas, M. Sim, The price of robustness. Oper. Res. **52**(1), 35–53 (2004)
22. A.J. Conejo, J.M. Morales, L. Baringo, Real-time demand response model. IEEE Trans. Smart Grid **1**(3), 236–242 (2010)
23. M. He, S. Murugesan, J. Zhang, Multiple timescale dispatch and scheduling for stochastic reliability in smart grids with wind generation integration, in *IEEE Conference on Computer Communications* (IEEE, 2011) pp. 461–465
24. C. Wu, H. Mohsenian-Rad, J. Huang, Wind power integration via aggregator-consumer coordination: A game theoretic approach, in *2012 IEEE Innovative Smart Grid Technologies* (IEEE, 2012) pp. 1–6
25. M.E. Khodayar, L. Wu, M. Shahidehpour, Hourly coordination of electric vehicle operation and volatile wind power generation in scuc. IEEE Trans. Smart Grid **3**(3), 1271–1279 (2012)
26. J. Kondoh, N. Lu, D.J. Hammerstrom, An evaluation of the water heater load potential for providing regulation service, in *Power and Energy Society General Meeting* (IEEE, 2011), pp. 1–8,

27. P. Du, N. Lu, Appliance commitment for household load scheduling. IEEE Trans. on Smart Grid **2**(2), 411–419 (2011)
28. D.E. Hoak, D.S. Parker, A.H. Hermelink, F.S.E. Center, How energy efficient are modern dishwashers?, in *Proceedings of the 2008 ACEEE Summer Study on Energy Efficiency in Buildings*, vol. 1, pp. 112–128, 2008
29. Commercial cooking appliance technology assessment (2013)

Chapter 4
Energy Generation Scheduling in Microgrids

4.1 Introduction

Energy generation scheduling to achieve robust and economic power supply is an essential component in microgrids. Two features of microgrids are the integration of large-scale renewable sources and the use of combined heat and power (CHP) generators. Such features, however, impose significant difficulties on the design of intelligent control strategies for microgrids. Traditional generation scheduling schemes are typically based on the perfect prediction of future demands [1], which is hardly the case in the microgrids since small-scale demands are hard to predict and renewable energies are highly volatile. Furthermore, although the integration of CHP generators can bring great economic benefits to microgrids by simultaneous production of useful thermal and power outputs, thereby increasing the overall efficiency and bringing environmental benefits, it brings new uncertainties to the scheduling problem: The heat demand exhibits a new stochastic pattern and makes it more difficult to predict the overall energy demands. On top of these, the real-time pricing in electricity market yields another uncertainty dimension to the scheduling problem. The microgrid has to make a proper strategic decision on the amount of power to import so as to cope with the financial risks brought by price uncertainty. Because of these unique challenges, it remains an open issue to design robust and cost-effective energy generation scheduling strategies for microgrids.

In this chapter, we consider a robust optimization-based energy generation scheduling problem in a CHP-microgrid scenario considering the net demand (the electricity demand not balanced by renewable energy) uncertainty, heat demand uncertainty, and electricity price uncertainty. The main contributions of this chapter can be briefly summarized as follows:

- We propose a new flexible uncertainty model to capture the fluctuant nature of the net demand and heat demand. Specifically, we extract reference distributions as useful references and allow the actual distributions of net demand and heat demand to vary around their references. To the best of our knowledge, this is the first time

© Springer Nature Singapore Pte Ltd. 2018

R. Wang et al., *Intelligent Microgrid Management and EV Control Under Uncertainties in Smart Grid*, https://doi.org/10.1007/978-981-10-4250-8_4

that distribution uncertainty model is adopted to depict the indeterminacy nature of net demand and heat demand.
- We develop chance-constraint approximation and robust optimization approaches based on our uncertainty model to transform the constraints with random variables into typical linear constraints. An iterative algorithm is designed to solve the problem.
- Price uncertainty is addressed by adopting robust optimization techniques, which allows the degree of conservatism to be controlled easily. We finally transformed the prime problem into a mixed-integer linear programming (MILP) problem, which can be solved efficiently by commercial solvers.
- The wide expandability of the proposed method is discussed, which shows its good applicable merits.
- Numerical results based on real-world data evaluate the impacts of different parameters and provide some suggestions on designing investment policies for microgrid. It is also shown that the proposed energy generation scheduling strategy achieves considerable cost savings, and the integration of CHP generators can effectively reduce the system expenditure.

The remainder of this chapter is organized as follows. Section 4.2 introduces the particulars of the system operation. In Sect. 4.3, we introduce the mathematical depiction of the energy generation scheduling problem and the uncertainty models of net and heat demands. Section 4.4 presents the chance-constraint approximation and robust optimization approach for handling the demand balancing and price uncertainty. In Sect. 4.5, several extensions of our proposed method are discussed. The simulation results and discussions are shown in Sect. 4.6. The parameters and calibration data are drawn from real-world statistics. Finally, we conclude this chapter in Sect. 4.7.

4.2 System Model

We consider a microgrid comprising a number of homogeneous CHP generators, a renewable energy generation system, and a local heating system. The microgrid is operated on the grid-connected mode, such that it can purchase electricity from the external utility grid when needed. The illustration of the microgrid system is shown in Fig. 4.1. The main symbols are utilized in this chapter, and their meanings are listed in Table 4.1. The particulars of the system operation are explained in the following subsections.

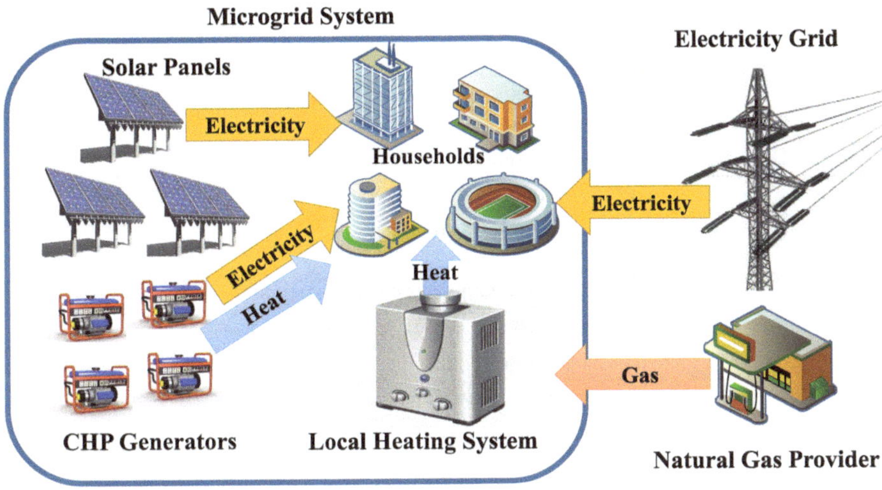

Fig. 4.1 An illustration of a typical microgrid system

4.2.1 CHP Generators

We divide time into discrete time slots with equal length. Let $\mathcal{A}$ denote the set of CHP generators. Further denote the start-up cost of turning on generator a as c_a^s, the sunk cost of maintaining the generator a in active mode for one time unit as c_a^b, and the marginal cost for generator a to produce one unit of electricity as c_a^m. Adopting a general generator model, we define energy generation scheduling vector $\mathbf{x_a}$ and state vector $\mathbf{y_a}$ as follows:

$$\mathbf{x_a} = [x_a^1, x_a^2, \ldots, x_a^H] \text{ and } \mathbf{y_a} = [y_a^1, y_a^2, \ldots, y_a^H], \tag{4.1}$$

where $H \geq 1$ is the scheduling horizon which indicates the number of time slots ahead that are taken into account for decision-making in the energy generation scheduling. For each coming time slot $h \in \mathcal{H} = [1, 2, \ldots, H]$, let a binary variable $y_a^h = 0/1$ denote the state of generator a (on/off) and a variable x_a^h denote the dispatched load to generator a. For each generator a with the maximum electricity output capacity E_a^{max} and the minimum stable generation E_a^{min}, we have

$$y_a^h \cdot E_a^{min} \leq x_a^h \leq y_a^h \cdot E_a^{max}. \tag{4.2}$$

The CHP generators can efficiently generate electricity and useful thermal energy simultaneously. Let η_a denote the heat–electricity ratio for generator a, which means that the CHP generator a can supply η_a units of heat for free when generating one unit of electricity. Alternatively, heat can be supplied by local heating system at a price of p_g per unit. We use the variable U^h to denote the amount of heat generated from

Table 4.1 Notations used in this Chapter 4

Symbol	Definition
$\mathcal{A}$	Set of CHP generators
a	Index of CHP generator, $a \in \mathcal{A}$
c_a^s	Start-up cost of turning on the generator a
c_a^b	Sunk cost of maintaining the generator a
c_a^m	Marginal cost for the generator a
$\mathcal{H}$	The set of time slots
$\mathbf{x_a}$	Energy generation scheduling vector of CHP a
$\mathbf{y_a}$	State vector of CHP a (binary)
E_a^{min}	The minimum stable output capacity of CHP a
E_a^{max}	The maximum electricity output capacity of CHP a
η_a	Heat–electricity ratio for the generator a
p_g	Price of heating system for providing one unit of heat
U^h	Amount of heat generated from heating system at time h
p_s^h	Electricity market price at time h
$\hat{p}_s^h$	Lower bound of the predicted electricity market price at time h
d^h	Uncertainty range of electricity market price at time h
V^h	Electricity obtained from outside power grid at time h
L^h	Net demand at time h (random variable)
S^h	Heat demand of the microgrid at time h (random variable)
$f_0(L^h)$	Electricity demand distribution at time h
$g_h(L^h)$	reference distribution of $f_0(L^h)$
D_h	Distance limit of $f_0(L^h)$'s uncertainty set
$U_r(\cdot)$	Uncertainty set based on KL divergence
ϵ	Fault tolerance limit of the power grid

local gas heaters at time slot h. Note that in this chapter, we omit the ramping-up and ramping-down constraints of CHP generators since we consider fast-response CHP generators such as gas turbines or microturbines, which have fast ramping rates and are able to start from cold to full capacity in 1–10 min [2].

4.2.2 *Electricity from External Utility Grid*

The microgrid can import electricity from outside electricity grid for the unbalanced power demand in an on-demand manner. We assume that the electricity market price at time h is p_s^h, which is a bounded random variable that takes value in $[\hat{p}_s^h, \hat{p}_s^h + d^h]$. $\hat{p}_s^h$ denotes the lower bound of the predicted price. $d^h > 0$ denotes that there exists price uncertainty (financial risks) at time h while $d^h = 0$ indicates the price at time h is known in advance. The amount of electricity obtained from electricity grid at time h is denoted as V^h.

4.2.3 Fluctuant Electricity and Heat Demand

Renewable energy generation can be regarded as a non-positive demand [1]. Denote the net demand at time h as L^h, which is a random variable of which the probability distribution may not be known. Similarly, the heat demand of the microgrid S^h is also random. Accurate prediction of small-scale demands and renewable energy generation is difficult to obtain due to limited management resources and their unpredictable nature. We need a proper uncertainty model to capture the indeterminacy properties of net and heat demands. A central requirement to the microgrid is to set the generation source power such that the electricity and heat supplies could meet the demands. This statement can be described as

$$V^h + \sum_{a \in \mathcal{A}} x_a^h \geq L^h \tag{4.3}$$

$$U^h + \sum_{a \in \mathcal{A}} \eta_a \cdot x_a^h \geq S^h. \tag{4.4}$$

4.3 Problem Formulation

In this section, a cost minimization problem formulation which incorporates CHP generation constraints, uncertain net demand, uncertain heat demand, and time-varying electricity prices is first given. The uncertainty model for describing the randomness of net demand and heat demand is then demonstrated.

4.3.1 Cost Minimization Formulation

The microgrid aims to minimize the operation cost of the whole system over the entire time horizon. The cost minimization formulation is defined as follows:

$$\begin{aligned} \min_{\mathbf{X},\mathbf{Y},\mathbf{V},\mathbf{U}} \quad & \sum_{h=1}^{H} \Bigg\{ p_g \cdot U^h + p_s^h \cdot V^h + \\ & \sum_{a \in \mathcal{A}} \left[c_a^m \cdot x_a^h + c_a^b \cdot y_a^h + c_a^s \cdot (y_a^h - y_a^{h-1})^+ \right] \Bigg\} \\ \text{s.t.} \quad & (4.2)\ (4.3)\ (4.4),\ y_a^h \in \{0, 1\} \\ & x_a^h, V^h, U^h \in \mathbb{R}_0^+, h \in \mathcal{H}, a \in \mathcal{A}, \end{aligned} \tag{4.5}$$

where $\mathbf{X} = [\mathbf{x_1}, \mathbf{x_2}, \ldots, \mathbf{x_a}, \ldots]^{\mathrm{T}}$ and $\mathbf{Y} = [\mathbf{y_1}, \mathbf{y_2}, \ldots, \mathbf{y_a}, \ldots]^{\mathrm{T}}$ are matrices of decision vectors $\mathbf{x_a}$ and $\mathbf{y_a}$ for $a \in \mathcal{A}$, respectively; $\mathbf{V} = [V^1, V^2, \ldots, V^h, \ldots]$ and

$\mathbf{U} = [U^1, U^2, \ldots, U^h, \ldots]$ are vectors of decision variables V^h and U^h for $h \in \mathcal{H}$, respectively; $(\cdot)^+$ is a function where $(x)^+ = \max(0, x)$. The cost function comprises the cost of electricity from outside power grid, the cost of generating heat from local heat generators, and the operation and start-up cost of CHP generators for the entire time horizon H.

A difficulty in solving this problem lies in the correlation term $(y_a^h - y_a^{h-1})^+$. However, if we introduce an auxiliary variable z_a^h into the problem formulation, then an equivalent expression can be obtained as:

$$\begin{aligned} \min_{\mathbf{X},\mathbf{Y},\mathbf{Z},\mathbf{V},\mathbf{U}} \quad & \sum_{h=1}^{H} \Big\{ p_g \cdot U^h + p_s^h \cdot V^h + \\ & \sum_{a \in \mathcal{A}} \left[c_a^m \cdot x_a^h + c_a^b \cdot y_a^h + c_a^s \cdot z_a^h \right] \Big\} \\ \text{s.t.} \quad & z_a^h \ge 0, \; z_a^h \ge y_a^h - y_a^{h-1} \\ & (4.2)\ (4.3)\ (4.4), \; y_a^h, z_a^h \in \{0, 1\} \\ & x_a^h, V^h, U^h \in \mathbb{R}_0^+, h \in \mathcal{H}, a \in \mathcal{A}, \end{aligned} \tag{4.6}$$

where $\mathbf{Z}_{|\mathcal{A}| \times H}$ is the matrix of auxiliary variable z_a^h for $a \in \mathcal{A}$, $h \in \mathcal{H}$. Another difficulty in solving problem (4.5) is the indeterminacy of net demand L^h and heat demand S^h existing in constraints (4.3) and (4.4). Note that optimizing over the space defined by (4.3) and (4.4) amounts to solving an optimization problem with potentially large or even infinite number of constraints. Obviously, this realization of uncertainties is intractable. Next, we develop a practical and flexible model to capture the uncertainties of L^h and S^h.

4.3.2 Probability Distribution Measure of Uncertainties

It is generally difficult to characterize the net demand and heat demand. In our optimization model, operations on the random variables L^h and S^h are cumbersome and computationally intractable. Moreover, in practice, we may not know the precise distributions of L^h and S^h. Solutions based on assumed distributions hence may not be justified. We usually measure the variability of a random variable using its variance or second moments which, however, may not provide sufficient details in describing the random variables. In this chapter, we extract a reference distribution, rather than moment statistics, from historical data that will capture the distribution properties. Since net demand and heat demand distributions may fluctuate over time and hard to be described in closed-form expressions, we will adopt empirical distributions as useful references and allow the actual distributions to fluctuate around them. For example, we may assume that the net demand distribution $f_0(L^h)$ is shifting around a known distribution $g_h(L^h)$, which can be obtained based on predictions and long-term field measurements. In the following part of this chapter, we only show the way to deal with random variable L^h. The method to tackle with random variable S^h is exactly the same.

The discrepancy between $f_0(L^h)$ and its reference $g_h(L^h)$ can be described by a probabilistic distance measure: the Kullback–Leibler (KL) divergence [3], which is a non-symmetric measure of the difference between two probability distributions. Name these two distributions as $f(L^h)$ and $g(L^h)$, respectively. Generally, one of the distributions, say, $f(L^h)$, represents the real distribution through precise modeling, while the reference $g(L^h)$ is a closed-form approximation based on the theoretical assumptions and simplifications. The definition of the KL divergence between two continuous distributions is given as follows:

$$D_{KL}(f(L^h), g(L^h)) = \int_{L^h \in S} [\ln f(L^h) - \ln g(L^h)] f(L^h) \mathrm{d}L^h, \tag{4.7}$$

where S is the integral domain. When distributions $f(L^h)$ and $g(L^h)$ are close to each other, the distance measure is close to zero. Adopting the KL divergence, we define the distribution uncertainty set as follows:

$$U_r(g(L^h), D_0) = \{f(L^h) \mid \mathbb{E}_f[\ln f(L^h) - \ln g(L^h)] \leq D_0\}, \tag{4.8}$$

where $D_0 > 0$ represents a distance limit which may be obtained from empirical data or real-time measurement. It indicates net demand's variation level. If the net demand is very volatile, we have less confidence on the reference distribution and thus may set a larger distance limit.

Considering the electricity demand distribution $f_0(L^h)$ with reference distribution $g_h(L^h)$ and distance limit D_h, we have the following constraints for electricity demand distribution $f_0(L^h)$:

$$\mathbb{E}_{f_0}[\ln f_0(L^h) - \ln g_h(L^h)] \leq D_h \tag{4.9}$$

$$\mathbb{E}_{f_0}[1] = 1. \tag{4.10}$$

With (4.9) and (4.10), we are now ready to transform the constraint (4.3) (similarly for (4.4)) to allow efficient solution of the problem (4.6).

Note that in the proposed approach, renewable energy is treated as a non-positive demand. We integrate user demand and renewable energy generation together and denote it as the net demand. The combined uncertainties from both user and supply sides are described by an uncertainty set as defined in (4.9) and (4.10).

The proposed model also allows some convenient extensions to include and handle more components in the microgrid systems. For example, to incorporate the reserve constraint into the proposed model, we only need to add the reserve constraints, which are linear functions, into the formulation (4.5) and then add a quadratic reserve cost into the objective function [4]. The new problem could still be transformed into a mixed-integer programming (MIP) problem, and the algorithm to be introduced in the next section can still be applied with virtually no change.

Remark Proper estimations of reference distribution and distance limit may be obtained by various methods (e.g., the Kernel Density Estimator [5, 6]), typically involving analyzing a large amount of historical data. Detailed discussions on such approaches, however, are beyond the scope of this book.

4.4 Optimization Algorithms

In this section, we present the optimization algorithms for solving problem (4.6). We first develop a robust approach for handling constraints (4.3) and (4.4), and then decompose (4.6) into a sub-problem and a main problem to allow easier solution. Finally, a robust approach to tackle the financial risk inducted by time-varying electricity market clearing prices is demonstrated.

4.4.1 Robust Approach for Constraints (4.3) and (4.4)

As shown in (4.3), the net demand balance can be expressed as $V^h + \sum_{a\in\mathcal{A}} x_a^h \geq L^h$. In practice, a decision criterion is to properly set decision $V^h + \sum_{a\in\mathcal{A}} x_a^h$ to allow good confidence that (4.3) is satisfied. To achieve that, we may introduce a small value ϵ to control the degree of conservatism and change the above expression into a chance constraint:

$$\mathbf{P}\left(L^h \geq V^h + \sum_{a\in\mathcal{A}} x_a^h\right) \leq \epsilon \tag{4.11}$$

where ϵ is the fault tolerance limit of the power grid, representing the acceptable probability that the desirable power supply is not attained. Then, we can have this expression that

$$\max_{f_0(L^h)\in U_r(g_h, D_h)} \mathbf{P}\left(L^h \geq V^h + \sum_{a\in\mathcal{A}} x_a^h\right) \leq \epsilon, \tag{4.12}$$

which is equivalent to:

$$\max_{f_0(L^h)\in U_r(g_h, D_h)} \int_{V^h+\sum_{a\in\mathcal{A}} x_a^h}^{+\infty} f_0(L^h)\mathrm{d}L^h \leq \epsilon. \tag{4.13}$$

Defining $\mathcal{L}^h = V^h + \sum_{a \in \mathcal{A}} x_a^h$ as the robust electricity supply (ES) decision, which equals the amount of electricity generated and imported at time slot h, we introduce an auxiliary function as follows:

$$h(L^h, \mathcal{L}^h) = \begin{cases} 0, & L^h \leq \mathcal{L}^h; \\ 1, & L^h > \mathcal{L}^h. \end{cases} \tag{4.14}$$

The left part of inequality (4.13) then can be formulated into an optimization problem:

$$\max_{f_0(L^h)} \int_0^{+\infty} h(L^h, \mathcal{L}^h) \cdot f_0(L^h) \mathrm{d}L^h \tag{4.15}$$

$$\text{s.t.} \quad \mathbb{E}_{f_0}[\ln f_0(L^h) - \ln g_h(L^h)] \leq D_h \tag{4.16}$$

$$\mathbb{E}_{f_0}[1] = 1 \tag{4.17}$$

Define $K_f^h(\mathcal{L}^h) = \max_{f_0(L^h) \in U_r(g_h, D_h)} \int_0^{+\infty} h(L^h, \mathcal{L}^h) \cdot f_0(L^h) \mathrm{d}L^h$ as the worst-case fault probability. We can then get a worst-case mapping $\mathcal{M}_{wc}^h$ which maps the robust ES decision $\mathcal{L}^h$ to $K_f^h(\mathcal{L}^h)$:

$$\mathcal{M}_{wc}^h : \quad \mathcal{L}^h \longrightarrow K_f^h(\mathcal{L}^h). \tag{4.18}$$

Note that, the degree of conservatism depends on the values of fault tolerance limit ϵ and the distance limit of uncertainty set D_h. When a less conservative control sequence is desired, we shall set a higher fault tolerance limit and a more lenient distance limit. A trade-off exists between the degree of conservation and the reliability of the decision-making.

4.4.2 Sub-Problem: Determine the Robust ES Decision Threshold

Since there exists a random variable L^h in the constraint, we cannot solve energy generation scheduling problem (4.6) directly. As aforementioned, we decompose the problem into a sub-problem and a main problem. The goal of the sub-problem is to determine the robust ES decision threshold $\mathcal{L}^{h*}$ so that the constraint (4.3) can be transformed into a solvable form.

Proposition 4.1 Problem (4.15)–(4.17) is a convex optimization problem.

The proof of this proposition is shown in Appendix-A. Through Proposition 4.1 and Slater's condition, we can see that strong duality holds for problem (4.15)–(4.17). Adopting the Lagrangian method, we can obtain the worst-case fault probability $K_f^h(\mathcal{L}^h)$ as follows:

$$K_f^h(\mathcal{L}^h) = \min_{\tau,\eta} \max_{f_0(L^h)} \mathbb{E}_{f_0}\Big[h(L^h, \mathcal{L}^h) \\ -\eta - \tau \ln\frac{f_0(L^h)}{g_h(L^h)}\Big] + \tau D_h + \eta,$$

where $\tau \geq 0$ and η are Lagrangian multipliers associated with constraints (4.16) and (4.17), respectively. Let

$$\mathcal{P}(L^h, f_0, \tau, \eta) = \mathbb{E}_{f_0}\left[h(L^h, \mathcal{L}^h) - \eta - \tau \ln\frac{f_0(L^h)}{g_h(L^h)}\right],$$

the derivative of $\mathcal{P}(\mathcal{L}^h, f_0, \tau, \eta)$ with respect to f_0 can be derived as

$$\frac{\partial \mathcal{P}}{\partial f_0} = \lim_{t\to 0}\frac{1}{t}\Big[\mathcal{P}\big(f_0(L^h) + t\cdot g_0(L^h)\big) - \mathcal{P}\big(f_0(L^h)\big)\Big]$$

$$= \int_0^{+\infty}\left(h(L^h, \mathcal{L}^h) - \tau \ln\frac{f_0(L^h)}{g_h(L^h)} - \eta - \tau\right) g_0(L^h) \mathrm{d}L^h.$$

Adopting the Karush–Kuhn–Tucker (KKT) optimality condition, we have

$$h(L^h, \mathcal{L}^h) - \tau \ln\frac{f_0(L^h)}{g_h(L^h)} - \eta - \tau = 0 \tag{4.19}$$

$$\int_0^{+\infty} f_0(L^h)\mathrm{d}L^h = 1 \tag{4.20}$$

$$\mathbb{E}\left[\ln\frac{f_0(L^h)}{g_h(L^h)}\right] - D_h \leq 0 \tag{4.21}$$

$$\tau\cdot\left(D_h - \mathbb{E}\left[\ln\frac{f_0(L^h)}{g_h(L^h)}\right]\right) = 0 \tag{4.22}$$

From (4.19), the optimal distribution function can be expressed as follows:

$$f_0^*(L^h) = g_h(L^h)\exp\left(\frac{h(L^h, \mathcal{L}^h) - \eta}{\tau} - 1\right). \tag{4.23}$$

The dual variables (τ, η) in (4.23) should be chosen properly such that conditions (4.20)–(4.22) are satisfied. Specifically, we have the following results.

Proposition 4.2 The choice of (τ, η) is a solution of the following nonlinear equations.

$$H_1(\tau, \eta) = R(\mathcal{L}^h)e^{-\eta/\tau} + S(\mathcal{L}^h)e^{(1-\eta)/\tau} - 1 = 0 \tag{4.24}$$

$$H_2(\tau, \eta) = S(\mathcal{L}^h)e^{(1-\eta)/\tau} - \eta - \tau(1 + D_h) = 0, \tag{4.25}$$

where $S(\mathcal{L}^h) = (1 - G_h(\mathcal{L}^h))\exp(-1)$, $R(\mathcal{L}^h) = G_h(\mathcal{L}^h)\exp(-1)$, and $G_h(\mathcal{L}^h) = \int_{L^h \leq \mathcal{L}^h} g_h(L^h)\mathrm{d}L^h$ denote the cumulative distribution function of reference distribution $g_h(L^h)$.

The proof for Proposition 4.2 is straightforward by substituting (4.23) to (4.20)–(4.22). However, it is still rather difficult to obtain an explicit solution from (4.24) and (4.25). Hence, we propose the Newton iteration method as detailed in Algorithm 4.1.

Once we determine the solutions for (4.24) and (4.25) in Proposition 4.2, we can obtain the worst-case fault probability from (4.19) and (4.22) as follows:

$$K_f^h(\mathcal{L}^h) = \mathbb{E}_{f_0^*}[h(L^h, \mathcal{L}^h)] = (1 + D_h)\tau + \eta. \tag{4.26}$$

Our next step is then to find the robust ES decision threshold $\mathcal{L}^{h*}$ such that $K_f^h(\mathcal{L}^{h*}) = \epsilon$, which involves the calculation of inverse function of $K_f^h(\mathcal{L}^h)$ that is not directly possible from (4.26). The following property of function $K_f^h(\mathcal{L}^h)$, however, may help us design such a search method.

Proposition 4.3 The worst-case fault probability $K_f^h(\mathcal{L}^h)$ is non-decreasing with respect to the robust ES decision $\mathcal{L}^h$.

It is straightforward to derive Proposition 4.3 since $\mathrm{d}K_f^h(\mathcal{L}^h)/\mathrm{d}\mathcal{L}^h = \mathrm{d}\mathbb{E}_{f_0^*}[h(L^h, \mathcal{L}^h)]/\mathrm{d}\mathcal{L}^h = f_0^*(\mathcal{L}^h) \geq 0$. Though direct solution is not available, the monotonicity of $K_f^h(\mathcal{L})$ enlightens us a bisection method to search for the solution for $K_f^h(\mathcal{L}^h) = \epsilon$. The main idea is to perform the search within an interval of $[0, \rho]$, where ρ is an empirical constant such that $K_f^h(\rho) > \epsilon$.

Details of the algorithm for searching the robust ES decision threshold are presented in Algorithm 4.1. Note that, from the 3rd to the 11th lines of the algorithm, we use Newton iteration to solve the equation in Proposition 4.2 and obtain the worst-case probability with fixed robust ES decision. Then, we compare the worst-case probability at $\mathcal{L}^h{}_-$ and $\mathcal{L}^{h-}$ with the fault tolerance limit ϵ, respectively. The comparison results help shrink the search region as shown in lines 12–14.

Once the robust ES decision threshold $\mathcal{L}^{h*}$ for the constraint (4.3) is obtained (and similarly, robust heat supply (HS) decision threshold $\mathcal{S}^{h*}$ for constraint (4.4) is obtained), we can approximate (4.3) and (4.4) with the following two constraints:

$$V^h + \sum_{a \in \mathcal{A}} x_a^h \geq \mathcal{L}^{h*} \tag{4.27}$$

$$U^h + \sum_{a \in \mathcal{A}} \eta_a \cdot x_a^h \geq \mathcal{S}^{h*}. \tag{4.28}$$

Algorithm 4.1 Search for robust ES decision threshold $\mathcal{L}^{h*}$

Input: Reference distribution $g_h(L^h)$;
Distance limit D_h; Search radius ρ;
Load balance fault tolerance limit ϵ; Tolerance ε.
Output: Robust ES decision threshold such that $K_f^h(\mathcal{L}^{h*}) = \epsilon$;
Begin
initialize $\mathcal{L}^h{}_- = 0$, $\mathcal{L}^{h^-} = \rho$, and set $\mathbf{H}(\tau, \eta) = [H_1(\tau, \eta), H_2(\tau, \eta)]^T$
while $|\mathcal{L}^h{}_- - \mathcal{L}^{h^-}| > \varepsilon$
  set $\bar{\mathcal{L}^h} = \frac{\mathcal{L}^h{}_- + \mathcal{L}^{h^-}}{2}$, initiate the time iteration $k = 1$
  while $\mathbf{H}(\tau, \eta) > \varepsilon$
    evaluate $\mathbf{H}(\tau, \eta)$ and Jacobian matrix $\mathbf{J}(\tau, \eta)$
    solve $\mathbf{J}(\tau, \eta)\Delta \mathbf{x_k} = -\mathbf{H}(\tau, \eta)$
    update $\tau_{k+1} = [\tau_k + \Delta\tau_k]^+$, $\eta_{k+1} = \eta_k + \Delta\eta_k$
    set $k = k + 1$
  end while
  update $K_f^h(\bar{\mathcal{L}^h}) = (1 + D_h)\tau_{k+1} + \eta_{k+1}$
  if $\left(K_f^h(\bar{\mathcal{L}^h}) - \epsilon\right)\left(K_f^h(\mathcal{L}^{h^-}) - \epsilon\right) < 0$
    then set $\mathcal{L}^h{}_- = \bar{\mathcal{L}^h}$ **else** set $\mathcal{L}^{h^-} = \bar{\mathcal{L}^h}$ **end if**
  if $|K_f^h(\bar{\mathcal{L}^h}) - \epsilon| < \varepsilon$ break **end if**
end while
set $\mathcal{L}^{h*} = \bar{\mathcal{L}^h}$
End

4.4.3 *Main Problem: Robust Approach for the Uncertain Electricity Prices*

There exist financial risks associated with real time electricity price uncertainty where p_s^h are unknown quantities. We adopt certain intervals at the $\alpha-$confidence level for prices $p_s^h \in [\hat{p}_s^h, \hat{p}_s^h + d^h]$, $h \in \mathcal{H}$ and formulate the well-defined robust model [7, 8]. Specifically, we tackle the following optimization problem rather than the original formulation (4.6):

$$\begin{aligned} \min \quad & \sum_{h=1}^{H} \Big\{ p_g \cdot U^h + \hat{p}_s^h \cdot V^h + \sum_{a \in \mathcal{A}} \Big[c_a^m \cdot x_a^h \\ & + c_a^b \cdot y_a^h + c_a^s \cdot z_a^h \Big] \Big\} + \phi \cdot \Gamma + \sum_{h \in J_0} e^h \\ \text{s.t.} \quad & \phi + e^h \geq d^h \cdot k^h, \quad \forall h \in J_0 \\ & -k^h \leq V^h \leq k^h \\ & e^h \geq 0, \quad k^h \geq 0, \quad \phi \geq 0, \quad z_a^h \geq 0, \quad \forall h \in J_0 \\ & z_a^h \geq y_a^h - y_a^{h-1} \\ & (4.2)\ (4.27)\ (4.28), \ y_a^h, z_a^h \in \{0, 1\} \end{aligned} \tag{4.29}$$

$$x_a^h, V^h, U^h, k^h, e^h, \Gamma \in \mathbb{R}_0^+, h \in \mathcal{H}, a \in \mathcal{A}.$$

Robust problem (4.29) is obtained using duality properties and exact linear equivalences. It represents the worse case while considering that electricity prices can be uncertain in at most Γ slots. $J_0 = \{h|\, d^h > 0\}$ is the set of electricity price $p_s^h, h \in \mathcal{H}$ that are subject to parameter uncertainty. Variable e^h is the dual variable of the initial problem (4.6) used to consider the known bounds of electricity prices, while ϕ and k^h are auxiliary variables used to obtain equivalent linear expression. Readers can refer to Appendix-B for detailed description of how to obtain this robust problem from problem (4.6). Γ is a parameter that controls the level of robustness in the objective function. This parameter is assumed to be integer and takes value in the set $\{0, 1, 2, \ldots, |J_0|\}$, i.e., between zero and the number of unknown electricity prices. In this case, when $\Gamma = 0$, the influence of price uncertainty in the objective function is ignored, while $\Gamma = |J_0|$, all possible price deviations are taken into account, which is the most conservative case. In general, a higher value of Γ increases the level of robustness at the expense of higher cost. Note that constraints (4.3) and (4.4) with random variables in the initial formulation (4.6) are approximated and replaced by (4.27) and (4.28) with no random variable. This problem is a mixed-integer linear programming (MILP) problem, which can be effectively tackled by cutting plane method, branch and bounded method, etc.

4.5 Possible Extensions of the Proposed Algorithm

With proper extensions, the proposed algorithm can be applied to handle some other scenarios regarding the energy generation scheduling problem in microgrids. A few possible extensions are briefly discussed as follows.

- The robust approach introduced in Sect. 4.4.1 also allows some convenient extensions to include and handle more constraints concerning the net demand (and the heat demand as well) uncertainty. For example, if the microgrid is confident that the density function $f_0(L^h)$ is not far from its reference $g_h(L^h)$ for any net demand, then the following l_∞ norm constraint with respect to $f_0(L^h)$ and $g_h(L^h)$ should be incorporated into problem (4.15)–(4.17):

$$\left\| f_0(L^h) - g_h(L^h) \right\|_\infty = \max\left\{ \left| f_0(L^h) - g_h(L^h) \right| \right\} \leq d^h, \qquad (4.30)$$

  where d^h is the maximum discrepancy between these two curve profiles. This constraint assures that there is no "sharp sting" protruding too far over the reference distribution curve. Since any norm is convex, the inequality constraint function on the left side of (4.30) is convex. Therefore, involvement of such a constraint will not influence the problem solving virtually. Algorithm 4.1 can still be adopted to obtain the robust ES/HS threshold with some minor revisions on $K_f^h(\mathcal{L}^h)$ and $\mathbf{H}(\tau, \eta)$, respectively.

- The microgrids' energy generation scheduling framework discussed in Sects. 4.3 and 4.4 is an offline approach suitable for planning the energy generation ahead of time. When real-time adjustment is of a big concern yet response time limit is not too rigid, the proposed algorithm can be easily extended to handle such cases. One option is to adopt the model predictive control (MPC) approach (also known as "receding horizon approach") [9, 10], of which the basic idea is to calculate the optimal control sequences yet implement only the first step of them. In other words, the energy scheduling problem is solved at time $h = \tau$ ($\tau \in \mathcal{H}$ denotes the current time index.) for the remaining horizon $[\tau, \tau+1, \ldots, H]$, yet only the solution for the current time slot τ is implemented. In the next time slot, the microgrid shall update the system information (e.g., the status of CHP generators, the net/heat demand prediction profiles, and robust ES/HS decision thresholds) and redo the calculations. For the cases where response time has to be very short, however, different algorithms with lower complexities and faster speed probably have to be developed, in order to support real-time operation and small-step scheduling more efficiently. The designing of such algorithms shall be considered in our future studies. Note that, for such cases, enhancing the system fault tolerance against the noises of real-time data shall also be considered.
- In this chapter, the discrepancy between net demand distribution $f_0(L^h)$ and its reference $g_h(L^h)$ is described by K–L divergence, which is a non-symmetric distance measure of the difference between two probability distributions. We select K–L divergence due to its wide applications. However, when the symmetric property of the measurement is of a big concern and some other forms of divergence (e.g., Jeffreys divergence[1] and squared Hellinger distance[2]) are adopted to characterize the distance between net demand distribution and its reference distribution, our method can be conveniently extended to tackle such cases. For instance, if the uncertainty set regarding the net demand distribution is defined through Jeffreys divergence, then the left part of inequality (4.13) can be reformulated into the following optimization problem:

$$\max_{f_0(L^h)} \int_0^{+\infty} h(L^h, \mathcal{L}^h) \cdot f_0(L^h) \mathrm{d}L^h \tag{4.31}$$

$$\text{s.t.} \int_0^{+\infty} \left(f_0(L^h) - g_h(L^h)\right) \left(\ln f_0(L^h) - \ln g_h(L^h)\right) \mathrm{d}L^h \le D_h \tag{4.32}$$

$$\int_0^{+\infty} f_0(L^h) \mathrm{d}L^h = 1. \tag{4.33}$$

---

[1]The Jeffreys divergence between distribution $p(x)$ and $q(x)$ is defined as $D_J(p||q) = \int(p(x) - q(x))(\ln p(x) - \ln q(x))\mathrm{d}x$.

[2]The squared Hellinger distance between distribution $p(x)$ and $q(x)$ is defined as $H^2(p||q) = 2\int(\sqrt{p(x)} - \sqrt{q(x)})^2\mathrm{d}x$.

Note that, compared with (4.15)–(4.17), the only difference of this formulation lies in the inequality constraint (4.32). However, considering the fact that the inequality constraint function in (4.32) is still convex with respect to $f_0(L^h)$, (4.31)–(4.33) still forms into a convex optimization problem. Therefore, the process to determine the robust ES decision threshold remains unchanged essentially. In order to obtain a new robust ES decision threshold, only $K_f^h(\mathcal{L}^h)$ and $\mathbf{H}(\tau, \eta)$ in Algorithm 4.1 need to be updated, and the algorithm can be applied with virtually no change. As for the squared Hellinger distance and many other forms of divergence, the situations are basically the same.

Note that, adopting the similar main idea of transforming a chance constraint with an unknown distribution into a solvable form, a different reformulation was proposed in [11], which provides a nice, closed-form solution for a special case where uncertainty set is defined through K–L divergence with no other constraint. In [11], the conjugate duality concept is adopted, and two decision variables (z and z_0) are optimized separately. Typically, optimizing over decision variables separately is not sufficient to reach global optimality. To ensure that such an approach can find the global optimal solution, a stringent requirement is that the optimal value of decision variable z_0^* can be expressed by z in closed form explicitly, which is suitable for K–L divergence case but is hard to be satisfied for other scenarios when other forms of divergence are adopted. Through our testing, for the special case (i.e., K–L divergence case), our method performs almost as good as the one in [11], yet our method has wider applications. As discussed above, our proposed method can conveniently handle uncertainty set defined by various measurement definitions, incorporating some other convex constraints.

4.6 Simulation Results and Discussions

In this section, we present simulation results based on real-world traces for assessing the performance of the proposed energy generation scheduling scheme and evaluate the effects of different parameters.

4.6.1 Parameters and Settings

4.6.1.1 Net Demand and Heat Demand Trace

We obtain the electricity and gas demand statistics from [12]. We focus on a college at Forecasting Climate Zone (FCZ) 09. The electricity within this zone is supplied by the Southern California Edison company. This trace contains hourly electricity demand and heat demand of the college in year 2002. We assume there are solar panels in the microgrid system. The area of solar panel in this microgrid system is set to be 3.75×10^4 m^2. The energy conversion efficiency is 0.8. The solar radiation intensity data is adopted from [13]. We employ electricity demand, heat demand, and solar power data of a typical month in winter (January) and estimate the distributions of net demand (electricity demand minus solar energy) and heat demand in each hour based on the samples using Kernel Density Estimation [14]. We find that in all the time slots (hours), the distribution functions of net demand and heat demand are close to be normal distribution. Thus, the reference distribution of net demand and heat demand is set to be normal distribution.

4.6.1.2 CHP Generator Characteristics

The parameters of CHP generators are set based on the statistics in [15]. The maximum output of a CHP generator is $E_a^{max} = 3.5$ MWh, and the minimum stable output is $E_a^{min} = 1.5$ MWh. The marginal cost for producing one unit of electricity is $c_a^m = 0.051$ \$/KWh, which is obtained using the fuel price and the energy conversion efficiency. The sunk cost for CHP generator keeping in active mode is $c_a^b = 110$ \$/h, which includes the capital cost, operation cost, and maintenance cost. We set the start-up cost to be $c_a^s = 560$ \$ and the heat–electricity ratio to be $\eta = 2.065$ [15]. Finally, it is assumed there are 8 CHP generators in this microgrid system unless otherwise stated.

4.6.1.3 Electricity and Gas Prices

The electricity price trace is obtained from [16], and the gas price data is obtained from [17]. In this chapter, we adopt the electricity market prices of central New York Control Area (NYCA) on a typical day in January. We set $\hat{p}_s^h$ and d^h equal to the lower bound and variation range of electricity market price at hour h. In addition, the natural gas price is set to be $p_g = 6.075$ \$/mmBTU.

4.6.2 *Results and Discussions*

4.6.2.1 Robust ES Threshold and Robust HS Threshold

We first solve the sub-problem and obtain the robust ES threshold $\mathcal{L}^{h*}$ and robust HS threshold $\mathcal{S}^{h*}$ for solving the main problem. The reference distributions of net demand and heat demand are normal and are estimated from sample data. The distance limit of net demand and heat demand uncertainty sets is 10^{-1}. The fault tolerance limit of net demand supply is 10^{-2} while the fault tolerance limit of heat demand supply is 10^{-1}. Given reference distributions, distance limits, and fault tolerance limits, we obtain $\mathcal{L}^{h*}$ and $\mathcal{S}^{h*}$ based on **Algorithm 4.1**. The results are shown in Table 4.2.

Table 4.2 Parameters of distribution uncertainty sets and corresponding ES and HS thresholds (unit: MWh for electricity and mmBTU for heat. $\overline{m}_E^h$ and σ_E^h are mean and standard deviation of net demand reference distribution, respectively. $\overline{m}_H^h$ and σ_H^h are mean and standard deviation of heat demand reference distribution, respectively)

Time slot	$\overline{m}_E^h$	σ_E^h	$\mathcal{L}^{h*}$	$\overline{m}_H^h$	σ_H^h	$\mathcal{S}^{h*}$
1	18.44	0.1059	18.98	63.88	8.3372	81.65
2	18.08	0.0965	18.57	51.96	5.0481	62.72
3	18.06	0.1005	18.58	43.63	1.7780	47.42
4	18.43	0.1246	19.07	46.62	1.8902	50.64
5	20.60	0.1456	21.34	50.39	1.7311	54.08
6	24.67	0.3807	26.61	80.35	7.5946	96.53
7	32.18	1.6355	40.52	124.93	1.4380	127.99
8	44.08	1.9485	53.50	283.69	8.0012	300.74
9	64.06	3.7971	78.77	285.91	6.4596	299.67
10	58.64	2.2394	60.54	254.82	7.5097	270.82
11	59.28	2.3199	58.88	219.39	10.7104	242.21
12	58.73	2.2730	56.27	195.55	10.1975	217.28
13	58.68	2.2121	54.18	183.64	11.0907	207.27
14	58.77	2.3731	56.66	177.02	11.6296	201.79
15	58.70	2.4761	61.38	171.43	12.0786	197.17
16	57.91	2.5475	64.65	167.69	12.1597	193.59
17	57.32	2.2805	67.77	166.47	12.6110	193.34
18	55.41	2.0156	65.69	169.83	14.0442	199.75
19	53.16	2.2647	64.72	176.10	14.0746	206.09
20	47.58	2.5553	60.62	184.35	14.3077	214.83
21	41.59	3.3157	58.51	190.49	15.3283	223.14
22	35.99	3.4268	53.47	198.32	15.0698	230.43
23	27.40	2.9277	42.34	111.43	10.2832	133.33
24	20.05	0.2638	21.40	78.80	7.7375	95.29

4.6.2.2 Potential Benefits of CHP Generators and Solar Panels

Once we obtain robust ES threshold $\mathcal{L}^{h*}$ and robust HS threshold $\mathcal{S}^{h*}$, we are ready to adopt the robust optimization approach to study the energy generation scheduling problem (4.29) with respect to real-time electricity market prices. Problem (4.29) is solved using the data provided in the previous subsection. The problem is solved using MOSEK optimization toolbox 7.0 on an Intel workstation with six processors clocking at 3.2 GHZ and 16 GB of RAM.

We first try to investigate the potential savings with CHP generators and solar panels. In particular, we conduct two sets of experiments. Both sets of experiments have nearly the same default settings, except that solar panels in the microgrid are enabled in the first set, but not in the second one. We vary the number of CHP generators installed in the microgrid from 0 to 10 and compute the total cost of the system in a day. The results are shown in Fig. 4.2. It is observed that 8 CHP generators with full capacity 28 MW are sufficient to obtain nearly all the cost-saving benefits. Thus, we may suggested that installed CHP generator capacity should be about half of the peak demand (The peak demand of a day in January is around 60 MW.). The intuitive reason is that most of the time demands are much lower than the peaks. This result can shed some light on making investment decisions in microgrids. Note that the leftmost points in the two curves denote the case where microgrid only uses external electricity and local heat generators (without CHP generators). System cost in this case can be interpreted as a cost benchmark. The results show that CHP can bring a saving of 6.2% (around \$5700 per day) to the system. Finally, by comparing the two curves in Fig. 4.2, we find that the one-day cost reduction achieved by solar panels is about 6.05% (around \$5200 per day).

4.6.2.3 Comparisons of Different Generation Scheduling Strategies

We compare three energy generation scheduling strategies: (1) the proposed robust optimal strategy (ROS); (2) fixed choice strategy (FCS): making one fixed choice of the generation level for entire duration for each generator. The system cost induced by this strategy has been used as a benchmark in literature [18]; (3) deterministic strategy (DS): A fixed number of CHP generators are switched on for the entire time horizon. The microgrid has to properly schedule the output level of active CHP generators, imported energy, and local heat generators to meet electricity and heat demand. Specifically, we consider three schemes with 0, 4, and 8 CHP generator(s) in active mode and termed as DS0, DS4, and DS8, respectively. We emphasize that the microgrid always tries to find the optimal control sequences under any of these three generation scheduling strategies, and the scheduling choices of the last two methods for comparison (i.e., FCS and DS) are made in hindsight. In addition, all the three scheduling strategies adopt the same parameter settings. The cost comparison results are depicted in Fig. 4.3.

As we observe in Fig. 4.3, ROS can achieve a cost saving of 4.5% (about \$3900 per day), 6.5% (about \$5700 per day), 1.2% (about \$1000 per day), and 5.0% (about \$4300 per day) compared with FCS, DS0, DS4, and DS8, respectively (equipped with solar panels). Moreover, we note that merely utilizing external electricity (DS0) or switching on all the local generators (DS8) are both not economical. Another interesting observation is that the cost of DS8 is lower than that of DS0. This shows that when all the CHP generators are switched on, although a significant amount of electricity may be wasted in the off-peak time slots, the strategy nevertheless still achieves better performance than the case where all electricity is imported from outside power grid. This justifies the economic potential of using local CHP generators. Obviously, DS4 achieves more cost savings than DS0 and DS8. This is because that when half of the CHP generators are turned on, a considerable proportion of the elec-

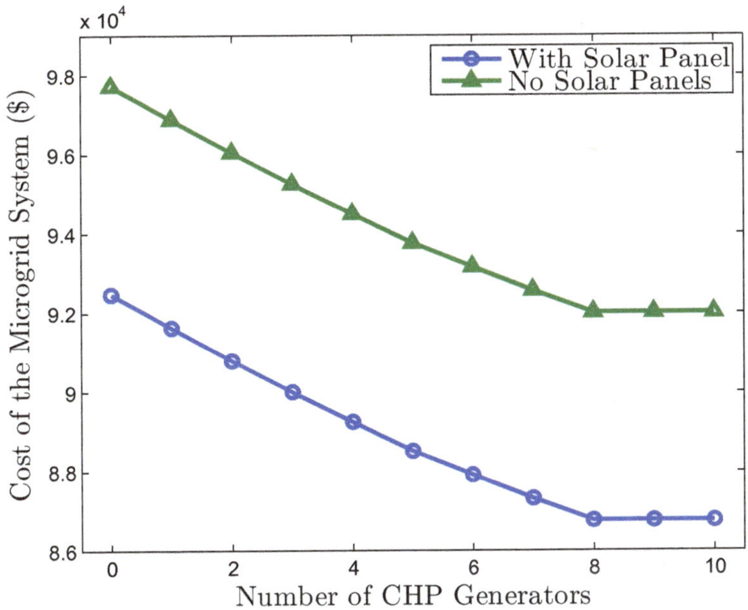

Fig. 4.2 Cost reduction for different number of CHPs

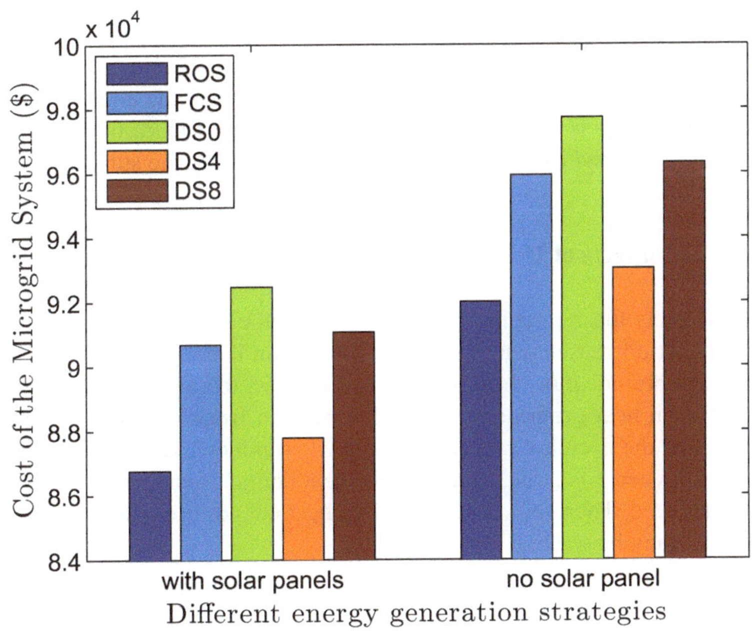

Fig. 4.3 Costs of different generation scheduling strategies

tricity demand can be supplied by CHP generators, and the energy loss in off-peak hours is relatively low than that in DS8.

4.6.2.4 The Impact of Robustness Level Γ

The sensitivities of the electricity cost with respect to robustness level Γ are depicted in Fig. 4.4. We set $|J_0| = 24$, i.e., price uncertainty may exist in all time slots of the day. We are interested in finding an optimal solution which optimizes against all scenarios under which a number Γ of the electricity price can vary in such a way as to maximally influence the objective. We vary the value of Γ from 0 to 24 in formulation (4.29) and obtain the optimal system cost. Remember that the value of Γ indicates the number of worst-case prices during the 24 time slots. $\Gamma = 0$ corresponds to the lowest robustness level while $\Gamma = 24$ corresponds to the highest robustness level. Apparently, the system cost is an increasing function of Γ. The incremental cost when the robustness level grows is the price for tackling the financial risks. We observe that to fully overcome the financial risks (i.e., the most conservatism condition), the microgrid has to pay additional 7.35% (about \$5900 per day) expenditures. However, the rise rate of the cost curve slows down when Γ increases. The reason is that when Γ increases, the protection level for the robust solution increases, then the probability that the robust solution is not favorable declines. Hence, it becomes less costly to protect the microgrid against the financial risk. We also compare the costs of two scenarios where solar panels are available and not available, respectively. The difference between these costs is called cost gap. It is interesting to note that cost gap only rises marginally when Γ increases. This shows that the uncertainty of solar energy has little impacts on the financial risks of the system since the indeterminacy of it has been alleviated by the proposed robust approach in the sub-problem.

4.6.2.5 The Impacts of Heat–Electricity Ratio η

Figure 4.5 depicts the reduction in cost versus heat–electricity ratio η. It appears that system cost decreases when η grows. The reason is that a larger η means CHP generators can provide more heat for free. In this case, the microgrid can reduce the reliance on local heat generators, which can be seen from Eq. (4.28). Meanwhile, we observe that the decrease rate slows down when η increases. This observation is intuitive since when η is large, nearly all the heat demands can be supplied by CHP generators for free. Therefore, additional free heat cannot bring significant benefits since the heat may be wasted.

4.6.2.6 System Cost Sensitivity to the Robust ES and HS Thresholds

In Fig. 4.6, we illustrate the relationship between the system cost and variation of $\mathcal{L}^{h*}$ and $\mathcal{S}^{h*}$. Specifically, we conduct two tests. In the first test, $\mathcal{S}^{h*}$ remains unchanged

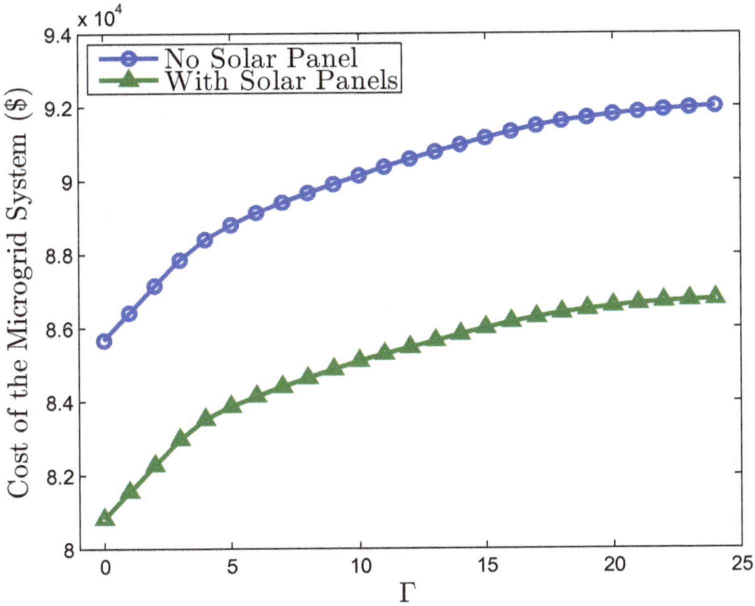

Fig. 4.4 System cost with respect to robustness level Γ

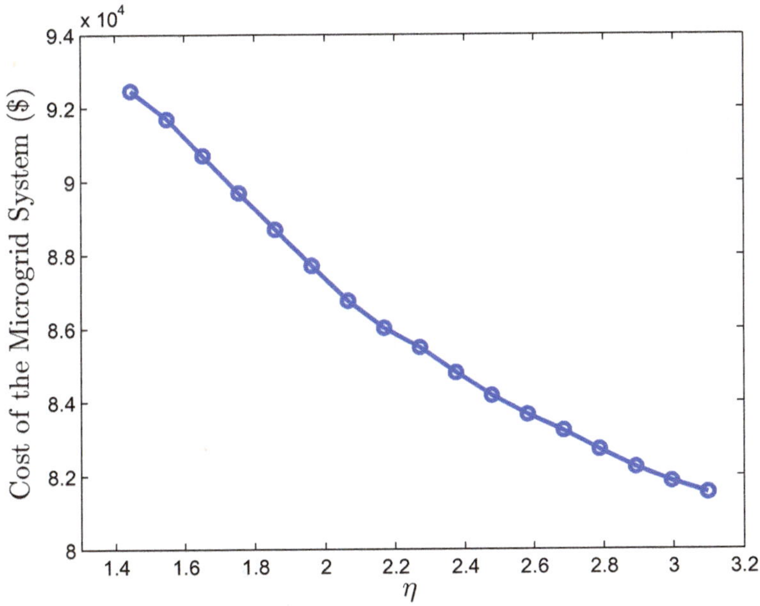

Fig. 4.5 Cost profile with respect to different η

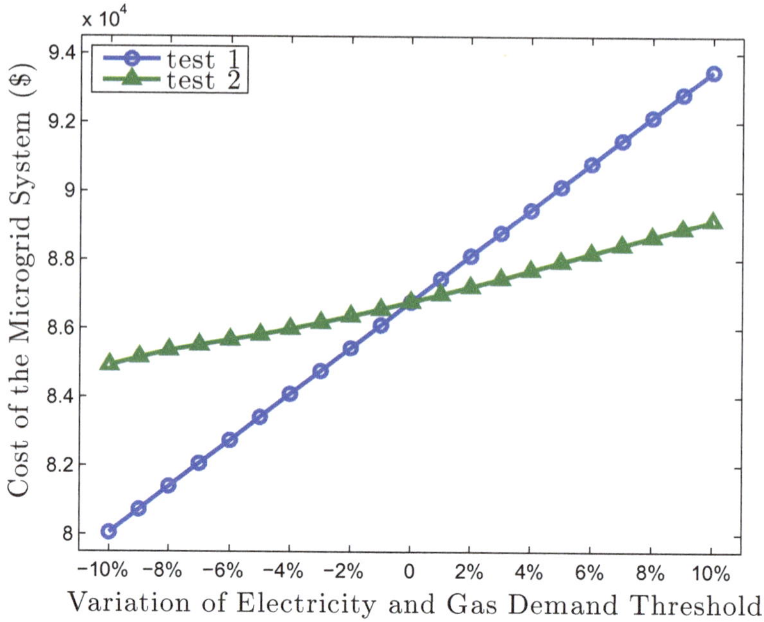

Fig. 4.6 Cost sensitivity with the variation of $\mathcal{L}^*$ and $\mathcal{S}^*$

and we vary the value of $\mathcal{L}^{h*}$; while in the second one, $\mathcal{L}^{h*}$ remains constant and $\mathcal{S}^{h*}$ varies. It is observed that the system cost has a nearly linear relationship with $\mathcal{L}^{h*}$ and $\mathcal{S}^{h*}$, which is consistent with the theoretical formulation (4.29). From (4.29), we see that the objective function have linear relationships with variables V^h, U^h, and $\sum_{a\in\mathcal{A}} x_a^h, h \in \mathcal{H}$. However, due to the trade-off between using local CHP generators and outside electricity when we vary $\mathcal{L}^{h*}$ and $\mathcal{S}^{h*}$, the relation between system cost and $\mathcal{L}^{h*}$ ($\mathcal{S}^{h*}$ as well) is only approximately linear. Also note that system cost is more sensitive to the variation of $\mathcal{L}^{h*}$. Since a large proportion of heat demands are satisfied by CHP generators for free, the system expenditure on heating is much lower than that on generating or buying electricity. Hence, the variation of heat demand has lower impacts on the system cost.

4.7 Conclusions

In this chapter, we study the energy generation scheduling problem in a microgrid scenario to minimize the cost and maintain system stability. To tackle the randomness of net demand and heat demand, we introduce reference distributions and then define distribution uncertainty sets to confine the fluctuations. Such a model allows convenient handling of volatile demands as long as the demand profiles are not too intensely different from the predictions or empirical knowledge. The uncertainty in electricity price is addressed by bounded random variables. We develop chance constraint approximations and robust optimization algorithms to firstly transform and then solve the problem. It is further shown that our proposed method can be conveniently extended to handle some other scenarios concerning the energy generation scheduling problem in microgrids. Numerical results based on real-world data indicate the satisfactory efficiency of the proposed energy scheduling strategy and the cost benefits of CHP generators. Moreover, the impacts of different parameters have been carefully evaluated. Such evaluations, as we believe, shall provide useful insights helping microgrid operators develop rational investment strategies.

References

1. C. Harris, *Electricity Markets: Pricing, Structures and Economics* (Wiley, New York, 2006)
2. Catalog of chp technologies (2008)
3. R.M. Gray, *Entropy and Information Theory* (Springer Science and Business Media, Berlin, 2011)
4. E. Afzalan, M. Joorabian, Emission, reserve and economic load dispatch problem with non-smooth and non-convex cost functions using epsilon-multi-objective genetic algorithm variable. Int. J. Electr. Power Energy Syst. **52**, 55–67 (2013)
5. E. Parzen, On estimation of a probability density function and mode. Ann. Math. Stat. 1065–1076 (1962)
6. M. Rosenblatt et al., Remarks on some nonparametric estimates of a density function. Ann. Math. Stat. **27**(3), 832–837 (1956)
7. D. Bertsimas, M. Sim, Robust discrete optimization and network flows. Math. Program. **98**(1), 49–71 (2003)
8. D. Bertsimas, M. Sim, The price of robustness. Oper. Res. **52**(1), 35–53 (2004)
9. E.F. Camacho, C.B. Alba, *Model Predictive Control* (Springer, London, 2004)
10. F. Allgöwer, A. Zheng, *Nonlinear Model Predictive Control* (Birkhäuser, Basel, 2000)
11. R. Jiang, Y. Guan, Data-driven chance constrained stochastic program (2013), http://www.optimization-online.org/DB_FILE/2013/09/4044.pdf
12. California commercial end-use survey (2013)
13. Nrel: National renewable energy laboratory (2013)
14. W. Zucchini, Kernel density estimation (2003)
15. Tecogen: Advanced modular chp systems (2012)
16. Nyiso: Market operation (2013)
17. Pg&e: Pacific gas and electric company (2013)
18. B. Narayanaswamy, V.K. Garg, T. Jayram, Online optimization for the smart (micro) grid, in *Proceedings of the International Conference on Future Energy Systems*, p. 19, ACM (2012)

Chapter 5
Energy Generation Scheduling in Microgrids Involving Temporal-Correlated Renewable Energy

5.1 Introduction

In typical energy generation scheduling problems, in order to tackle with the time-varying renewable energy and user demand, time is usually slotted and, in each time slot, the renewable energy and user demand are considered as either deterministic or random variables. Current literature rarely takes the correlation information of renewable energy generation or user demand into account when they design the energy generation scheduling schemes. Considering the fact that in practical scenario, renewable energy generations or user demands are correlated over time [1, 2], we aim at evaluating how such information influences the performance of the optimal energy generation scheduling strategy. If the correlation information can improve the performance of the scheme to a large margin, then utilizing such information is valuable. For this problem, mathematical tools such as the generalized Gaussian inequality framework and the semidefinite programming (SDP) may help us investigate and solve it.

In this chapter, by extending the previous work in Chap. 4, we further consider a robust optimization-based energy generation scheduling problem in a microgrid scenario considering the uncertainty of renewable energy and integration of energy storages. The main contributions of this chapter can be briefly summarized as follows:

- We adopt the moment statistic model to capture the fluctuant nature of the renewable energy. To the best of our knowledge, this is the first time that moment statistics are utilized to model and analyze the properties of renewable energy generation. In addition, moment statistics are easy to obtain in practical applications. Compared with the distribution uncertainty model proposed in our previous work [3, 4], the microgrid systems do not need to analyze a large amount of historical data by adopting the moment statistic model.
- The energy generation scheduling problem is formulated into a cost minimization problem with random variables in the constraints. We develop chance-constraint approximations and robust optimization approaches to transform the problem into a solvable form.

© Springer Nature Singapore Pte Ltd. 2018

R. Wang et al., *Intelligent Microgrid Management and EV Control Under Uncertainties in Smart Grid*, https://doi.org/10.1007/978-981-10-4250-8_5

- To the best of our knowledge, this work is the first to investigate how the temporal-correlation information of renewable energy impacts the energy generation scheduling in microgrids.
- Numerical results based on real-world data evaluate the impacts of different parameters and performance bounds of the proposed scheduling scheme. A novel observation is that the temporal-correlation information of the renewable energy can help to effectively reduce the conservativeness of the problem solving and improve the performance of the proposed generation scheduling scheme.

The rest of this chapter is organized as follows. Section 5.2 introduces the particulars of the system operation. In Sect. 5.3, we introduce the mathematical depiction of the energy generation scheduling problem and the moment statistic model of renewable energies. Section 5.4 presents the chance-constraint approximation and robust optimization approach for handling the demand balancing and renewable energy uncertainties. The simulation results and discussions are shown in Sect. 5.5. The parameters and calibration data are drawn from real-world statistics. Finally, we conclude our this chapter in Sect. 5.6.

5.2 System Model

We consider a microgrid comprising a number of homogeneous conventional power units, a renewable energy generation system (e.g., solar panels), and an energy storage system. Currently, the microgrid is operated in the islanded mode. The illustration of the microgrid system is shown in Fig. 5.1. The particulars of the system operation are explained in the followings.

We divide time into discrete time slots with an equal length. Let $\mathcal{A}$ denote the set of conventional power generators. Further denote the start-up cost for turning on a generator a as c_a^s, the sunk cost of maintaining the generator a in active mode for one unit of time as c_a^b, and the marginal cost for the generator a to produce one unit of electricity as c_a^m. Adopting a general power unit model, we define the energy generation scheduling vector $\mathbf{x_a}$ and state vector $\mathbf{y_a}$ as follows:

$$\mathbf{x_a} = [x_a^1, x_a^2, \ldots, x_a^H] \text{ and } \mathbf{y_a} = [y_a^1, y_a^2, \ldots, y_a^H], \tag{5.1}$$

where $H \geq 1$ is the scheduling horizon which indicates the number of time slots ahead that are taken into account for decision-making in the energy generation scheduling. For each coming time slot $h \in \mathcal{H} = [1, 2, \ldots, H]$, we use a binary variable $y_a^h = 0/1$ to denote the state of generator a (off/on) and a variable x_a^h to denote the dispatched load to power unit a. For each unit a with a maximum power output capacity E_a^{max} and a minimum stable output E_a^{min}, we have

$$y_a^h \cdot E_a^{min} \leq x_a^h \leq y_a^h \cdot E_a^{max}. \tag{5.2}$$

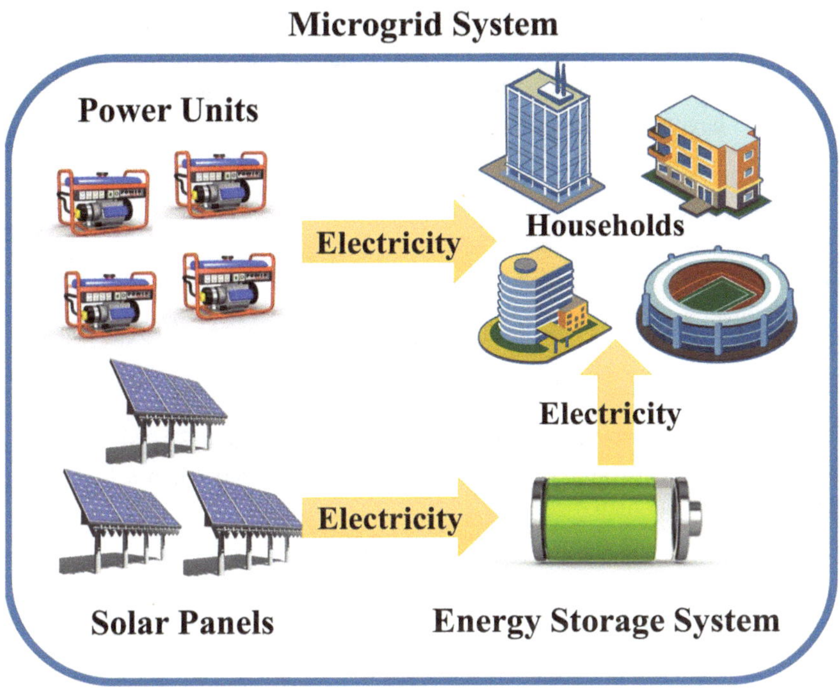

Fig. 5.1 Architecture of a typical microgrid system

In our model, we assume that the renewable energy harvested from solar panels will be first saved into energy storage devices for future use; i.e., a solar-plus-battery system is considered. The household can obtain electricity from energy storages in an on-demand manner. Denote the household demand and energy obtained from energy storages at time h as D^h and V^h, respectively. A central requirement of the microgrid is to set the energy source power such that the electricity could meet the demand at all time slots. This statement can be described as

$$\sum_{h=1}^{H} x_a^h + V^h = D^h, \ \forall h \in \mathcal{H}. \tag{5.3}$$

Let B^h denote the amount of energy stored in the battery. To ensure that there is always backup power for emergency use, we require the battery to be maintained at or above its initial level at the end of the scheduling horizon:

$$\sum_{h=1}^{H} V^h - \sum_{h=1}^{H} \xi^h \leq 0, \tag{5.4}$$

where $\xi^h \in [0, \xi^{max}]$ is the random variable representing the amount of energy harvested from renewable energy devices (e.g., solar panels), and ξ^{max} denotes the maximum generation capacity of the renewable energy generators. A battery's level can never go beyond the maximum capacity or drop below 0. Therefore, we have that

$$0 \leq B^h \leq B^{max}, \tag{5.5}$$

where B^{max} represents the maximum capacity of the energy storage devices. Last, the battery level varies over time as

$$B^{h+1} = B^h + \xi^h - V^h. \tag{5.6}$$

In this chapter, we assume that the energy storage device is of a large size. Under such case, constraints (5.5) and (5.6) can be relaxed when scheduling the energy generation in microgrids [5].

5.3 Problem Formulation

5.3.1 *Cost Minimization Formulation*

The microgrid aims to minimize the operation cost of the whole system over the entire time horizon. The cost minimization formulation is defined as follows

$$\begin{aligned} \min_{\mathbf{X},\mathbf{Y},\mathbf{V}} \quad & \sum_{h=1}^{H} \sum_{a \in \mathcal{A}} \left[c_a^m \cdot x_a^h + c_a^b \cdot y_a^h + c_a^s \cdot (y_a^h - y_a^{h-1})^+ \right] \\ \text{s.t.} \quad & (5.2) - (5.4), \ y_a^h \in \{0, 1\} \\ & x_a^h, V^h \in \mathbb{R}_0^+, h \in \mathcal{H}, a \in \mathcal{A}, \end{aligned} \tag{5.7}$$

where $\mathbf{X} = [\mathbf{x_1}, \mathbf{x_2}, ..., \mathbf{x_a}, ...]^{\mathrm{T}}$ and $\mathbf{Y} = [\mathbf{y_1}, \mathbf{y_2}, ..., \mathbf{y_a}, ...]^{\mathrm{T}}$ are matrices of decision vectors $\mathbf{x_a}$ and $\mathbf{y_a}$ for $a \in \mathcal{A}$, respectively; $\mathbf{V} = [V^1, V^2, ..., V^h, ...]$ is the vector of decision variables V^h for $h \in \mathcal{H}$; $(\cdot)^+$ is a function where $(x)^+ = \max(0, x)$. The cost function comprises the operation and start-up costs of conventional power generators for the entire time horizon H.

A difficulty in solving this problem lies in the correlation term $(y_a^h - y_a^{h-1})^+$. By introducing an auxiliary variable z_a^h into the problem formulation, an equivalent expression can be obtained as

$$\min_{\mathbf{X},\mathbf{Y},\mathbf{Z},\mathbf{V}} \sum_{h=1}^{H}\sum_{a\in\mathcal{A}}\left[c_a^m \cdot x_a^h + c_a^b \cdot y_a^h + c_a^s \cdot z_a^h\right] \tag{5.8}$$
$$\text{s.t.} \quad z_a^h \geq 0,\ z_a^h \geq y_a^h - y_a^{h-1}$$
$$(5.2) - (5.4),\ y_a^h, z_a^h \in \{0,1\}$$
$$x_a^h,\ V^h \in \mathbb{R}_0^+, h \in \mathcal{H}, a \in \mathcal{A},$$

where $\mathbf{Z}_{|\mathcal{A}|\times H}$ is the matrix of auxiliary variable z_a^h for $a \in \mathcal{A}$, $h \in \mathcal{H}$. The objective for introducing an auxiliary variable z_a^h into problem formulation (5.7) is to have an equivalent, solvable problem without the correlation term $(y_a^h - y_a^{h-1})^+$. Another difficulty in solving problem (5.7) is the indeterminacy of renewable energy generations ξ^h existing in (5.4). Note that to optimize over the space defined by (5.4) amounts to solving an optimization problem with potentially large or even infinite number of constraints. Obviously, this realization of uncertainties is intractable. Next, we adopt the moment statistic model to capture the uncertainties of ξ^h.

5.3.2 *Moment Statistic Model*

It is generally difficult to characterize the renewable energy generation. However, we may measure the variability of renewable energy generation using its mean and second-order moments, which are quite easy to obtain from field measurements. Mathematically, we may assume that renewable energy generation $\xi = [\xi^1, ..., \xi^H]$ is confined by the following uncertainty set:

$$\mathcal{P}(\mu, S) = \left\{\mathbb{P}_\xi \in \mathcal{P}_\infty : \int_{\mathbb{R}^n} \xi \cdot \mathbb{P}(d\xi) = \mu, \right. \tag{5.9}$$
$$\left. \int_{\mathbb{R}^n} \xi\xi^T \mathbb{P}(d\xi) = S\right\},$$

where $\mu \in \mathbb{R}^H$ and $S \in \mathbb{S}^H$, $\mathbb{S}^H$ is the set of symmetric matrixes with dimension H, while $\mathcal{P}_\infty$ represents the set of all distributions on $\mathbb{R}^H$. Thus, $\mathcal{P}(\mu, S)$ contains all distributions that share the same mean μ and second-order moment matrix S. The temporal-correlated information of renewable energy generation is included in S; e.g., the two lines above and below the diagonal of S indicate the correlation within one time slot. With this moment statistic model, we are now ready to transform the constraint (5.4) to allow efficient solution of (5.8).

5.4 Optimization Algorithm

5.4.1 Robust Approach for Constraint (5.4)

As shown in (5.4), the energy storage balance can be expressed as: $\sum_{h=1}^{H} V^h - \sum_{h=1}^{H} \xi^h \leq 0$. In practice, a decision criterion is to properly set decision vector $\mathbf{V}$ to allow good confidence that (5.4) is satisfied. To achieve that, we may introduce a small value ϵ to control the degree of conservativeness and change the above expression into a chance constraint

$$\mathbf{P}\left(\sum_{h=1}^{H} \xi^h < \sum_{h=1}^{H} V^h\right) \leq \epsilon, \tag{5.10}$$

where ϵ is the fault tolerance limit of the microgrid, representing the acceptable probability that the desirable power supply is not attained. Then we can have the robust expression that

$$\sup_{\mathbb{P}_\xi \in \mathcal{P}(\mu,S)} \mathbf{P}\left(\sum_{h=1}^{H} \xi^h < \sum_{h=1}^{H} V^h\right) \leq \epsilon. \tag{5.11}$$

Theorem 1 *Solving the left part of inequality (5.11) is equivalent to solving the following semidefinite programming problem (SDP):*

$$\begin{aligned} \max \ & \sum_{i=1}^{k} \lambda_i \\ s.t. \ & z_i \in \mathbb{R}^H, Z_i \in \mathbb{S}^H, \lambda_i \in \mathbb{R} \quad \forall i = 1, 2, \ldots, k \\ & a_i^T z_i \geq b_i \lambda_i \quad \forall i = 1, 2, \ldots, k \\ & \sum_{i=1}^{k} \begin{pmatrix} Z_i & z_i \\ z_i^T & \lambda_i \end{pmatrix} \preceq \begin{pmatrix} S & \mu \\ \mu^T & 1 \end{pmatrix} \\ & \begin{pmatrix} Z_i & z_i \\ z_i^T & \lambda_i \end{pmatrix} \succeq 0 \quad \forall i = 1, 2, \ldots, k \end{aligned} \tag{5.12}$$

where $a_{1_{H\times 1}} = -1 \cdot [1, 1, \ldots, 1]^T$; $[a_2, \ldots, a_{H+1}] = -1 \cdot \mathbf{I}_H$; $[a_{H+2}, \ldots, a_{2H+1}] = \mathbf{I}_H$, *and* $\mathbf{I}_H$ *is the identity matrix with dimension* H; $b_1 = \sum_{h=1}^{H} V^h$; $[b_2, \ldots, b_{H+1}] = [0, \ldots, 0]^T$; $[b_{H+2}, \ldots, b_{2H+1}] = \xi^{max} \cdot [1, 1, \ldots, 1]^T$, *and obviously* $k = 2H + 1$.

The SDP reformulation (5.12) can be obtained through the generalized Chebyshev inequality bounds. Detailed proof of **Theorem** 1 is lengthy and omitted here due to limited space. Readers may refer to reference [6] for more detailed descriptions.

Defining $b_1 = \sum_{h=1}^{H} V^h$ as the robust electricity acquisition (EA) decision, which is equal to the amount of electricity obtained from energy storage systems during the whole time horizon. Further define $K_\xi(b_1) = \sup_{\mathbb{P}_\xi \in \mathcal{P}(\mu,S)} \mathbf{P}(\sum_{h=1}^{H} \xi^h < \sum_{h=1}^{H} V^h)$ as the worst-case fault probability. We can then get a worst-case mapping $\mathcal{M}_{wc}$ which maps the robust EA decision b_1 to $K_\xi(b_1)$:

$$\mathcal{M}_{wc}: \quad b_1 \longrightarrow K_\xi(b_1). \tag{5.13}$$

5.4.2 Determine the Robust EA Decision Threshold

Since there exist random variables in the constraint (5.4), we cannot solve energy generation scheduling problem (5.8) directly. As mentioned before, we adopt chance-constraint approximations and robust approaches to transform the constraint (5.4). The goal of such transformation is to determine the maximum robust EA decision b_1^* (i.e., robust EA decision threshold) so that the constraint (5.4) can be transformed into a solvable form.

Theorem 2 *The worst-case fault probability $K_\xi(b_1)$ is non-decreasing with respect to the robust EA decision b_1.*

It is straightforward to derive **Theorem** 2 since $\mathrm{d}K_\xi(b_1)/\mathrm{d}b_1 = f_\xi(b_1) \geq 0$, where f_ξ is the probability density function of random variable $\sum_{h=1}^{H} \xi^h$. Though directly obtaining the robust decision threshold is not practical, the monotonicity of $K_\xi(b_1)$ enlightens us a bisection method to search for the solution for $K_\xi(b_1^*) = \epsilon$. The main idea is to perform the search within an interval of $[0, \rho]$, where ρ is an empirical constant such that $K_\xi(\rho) > \epsilon$.

Details of the algorithm for searching the robust EA decision threshold are presented in Algorithm 1. Note that, in the fifth line of the algorithm, we use interior point method to solve the SDP problem in **Theorem** 1 and obtain the worst-case probability with fixed robust EA decision. Then we compare the worst-case fault probability at $b_{1\_}$ and $b_1{}^-$ with the fault tolerance limit ϵ, respectively. The comparison results help shrink the search region as shown in lines 6–9.

Once the robust EA decision threshold b_1^* for the constraint (5.4) is obtained, we can approximate (5.4) with the following constraint:

$$\sum_{h=1}^{H} V^h = b_1^*. \tag{5.14}$$

Now we can tackle the following optimization problem rather than the original formulation (5.8)

Algorithm 5.1 Search for robust EA decision threshold b_1^*

Input: Mean vector μ; Second-order moment matrix S
Search radius ρ; Battery balance fault tolerant limit ϵ;
Computational accuracy tolerance ϵ.
Output: Robust EA decision threshold such that $K_\xi(b_1^*) = \epsilon$;
Begin
initialize $b_{1\_} = 0, b_1^{-} = \rho$
while $|b_{1\_} - b_1^{-}| > \epsilon$
  set $\bar{b_1} = \frac{b_{1\_} + b_1^{-}}{2}$
  compute $K_\xi(\bar{b_1})$ by solving the SDP problem (5.12)
  if $\Big(K_\xi(\bar{b_1}) - \epsilon\Big)\Big(K_\xi(b_1^{-}) - \epsilon\Big) < 0$
    then set $b_{1\_} = \bar{b_1}$
  else set $b_1^{-} = \bar{b_1}$ **end if**
  if $|K_\xi(\bar{b_1}) - \epsilon| < \epsilon$ **break end if**
end while
set $b_1^* = \bar{b_1}$
End

$$\begin{aligned} \min_{\mathbf{X},\mathbf{Y},\mathbf{Z},\mathbf{V}} \quad & \sum_{h=1}^{H} \sum_{a \in \mathcal{A}} \left[c_a^m \cdot x_a^h + c_a^b \cdot y_a^h + c_a^s \cdot z_a^h \right] \\ \text{s.t.} \quad & z_a^h \geq 0,\ z_a^h \geq y_a^h - y_a^{h-1} \\ & (5.2)\ (5.3)\ (5.14),\ y_a^h, z_a^h \in \{0, 1\} \\ & x_a^h, V^h \in \mathbb{R}_0^+, h \in \mathcal{H}, a \in \mathcal{A}. \end{aligned} \tag{5.15}$$

Note that constraint (5.4) with random variables in the initial formulation (5.8) is approximated and replaced by (5.14) with no random variable. This problem is a mixed-integer linear programming (MILP) problem, which can be solved effectively by cutting plane method, branch and bounded method, etc.

5.5 Performance Evaluation and Analysis

In this section, we present numerical results based on real-world traces to assess the performance bounds of the proposed energy generation scheduling scheme and evaluate the effects of different parameters.

5.5.1 Parameters and Settings

We assume there are solar panels in the microgrid system. The area of solar panels in this microgrid system is set to be 1.5×10^4 m^2. The energy conversion efficiency is 0.4. The monthly clearness index time series are from 10 meteorological stations in Singapore. These stations are designed to perform monitoring of solar radiation. Silicon sensors are employed at each station, with some also having pyranometers that measure diffuse and global irradiance. The silicon sensors are calibrated by the Fraunhofer Institute for Solar Energy Systems to achieve an uncertainty under 2%. The data used in this work is hourly data collected by these 10 stations in November 2012 [1, 7].

We obtain the electricity demand statistics from [8]. We focus on a college at Forecasting Climate Zone (FCZ) 09. This trace contains hourly electricity demand of the college in year 2002. The parameters of conventional power generators are set based on the statistics in [9]. The maximum output of a power unit is $E_a^{max} = 3.5$ MWh, and the minimum stable output is $E_a^{min} = 1.5$ MWh. The marginal cost for producing one unit of electricity is $c_a^m = 0.051$ \$/KWh, which is obtained using the fuel price and the energy conversion efficiency. The sunk cost for a generator keeping in active mode is $c_a^b = 110$ \$/h, which includes the operation cost, capital cost, and maintenance cost. The start-up cost is set to be $c_a^s = 560$ \$. Finally, unless otherwise stated, it is assumed there are 10 power units in this microgird system, the duration of a time slot is 1 h, and the time horizon is 12 h. The MILP problem is solved using Mosek optimization toolbox 7.0 on an Intel workstation with six processors clocking at 3.2 GHZ and 16 GB of RAM.

5.5.2 Results and Discussion

We first investigate the statistical properties of solar energy generation in the time domain. In particular, we adopt solar irradiance data for the first two weeks in November 2012. Nonlinear least square method is used to obtain the fitted line. The results concerning the temporal coherence of solar energy generation with respect to time lag is depicted in Fig. 5.2. Note that the colored dots show the coherence of solar energy generations in 10 stations, and the blue curve is the fitted function $r_{tc} = 1 - 0.1644\tau + 0.0.0038\tau^2$, where τ is the time lag and r_{tc} is the coherence. As we observe in the figure, solar energy generations show near-linear correlation in the time domain, and such observations help us analyze the performance bounds of the proposed energy generation scheduling scheme in the following contents.

Next, we investigate how the robust EA decision threshold b_1^* varies when the fault tolerant limit ϵ increases. Figure 5.3 plots the mapping from fault tolerance limit ϵ to robust EA decision threshold b_1^*. It is shown from the figure that the robust EA decision threshold b_1^* grows when ϵ increases. In other words, a larger fault tolerant limit ϵ permits a higher reliance on the solar energy (a larger robust EA decision threshold), which is straightforward to understand. Note that the robust EA decision threshold function is monotone; therefore, it is justified to adopt the bisection method as presented in Algorithm 5.1 to search for the robust EA decision threshold. We also observe that the incremental rate of the robust EA decision threshold slows down when ϵ increases.

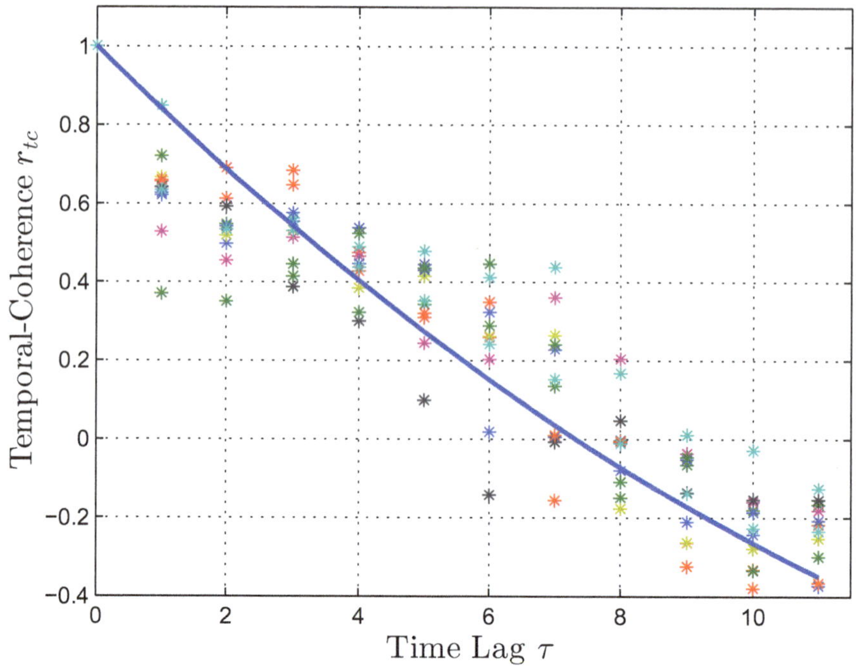

Fig. 5.2 Temporal-correlation fitting using first two week's radiation data of November 2012. Nonlinear least square method is adopted to get the fitted curve $r_{tc} = 1 - 0.1644\tau + 0.0.0038\tau^2$

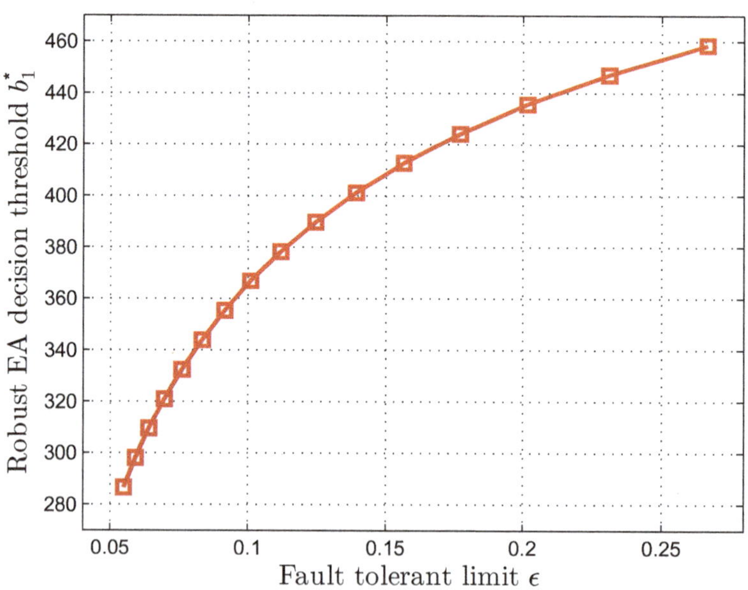

Fig. 5.3 Robust EA decision threshold b_1^* with respect to the fault tolerant limit ϵ

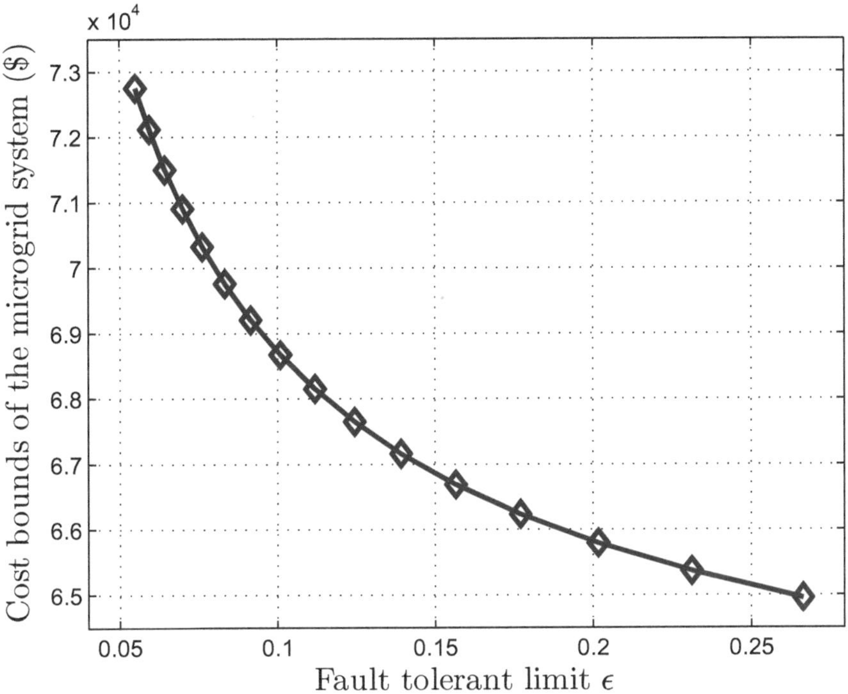

Fig. 5.4 Cost bounds of the microgrid system with respect to the fault tolerant limit ϵ

In Fig. 5.4, we vary the values of fault tolerance limit ϵ and study how system cost bound changes with respect to ϵ. Note that the cost bound represents the operation cost of the microgrid system under the worst-case condition of solar energy generation. Apparently, the cost bound decreases when ϵ increases. The reason is that when ϵ increases, the protection level for the robust solution will decrease, and the scheduling strategies of the microgrid hence become less conservative, leading to the decline of the operation cost. Also note that the cost bound is less sensitive when the fault tolerance limit ϵ is at a higher level.

In Fig. 5.5, we evaluate how the fault probability $K_\xi(b_1)$ varies with respect to the amount of temporal-correlation information utilized under different values of robust EA decision b_1. Specifically, we conduct a set of experiments. In the first step, we only utilize the mean and variance information of renewable energy generation to define μ vector and S matrix; i.e., the elements in S is set to 0 except for those on the diagonal; at the second step, we add the temporal-correlation information within 1-h time lag, i.e., only the diagonal, and the lines above and below it have values in the second-order moment matrix S. Then, at each time step n, $n \geq 2$, the temporal-correlation information within time lag 0 and time lag $n-1$ is utilized for the decision-making. We repeat such process until time step 12. At each step of the experiment, we compute the fault probability under different values of robust EA

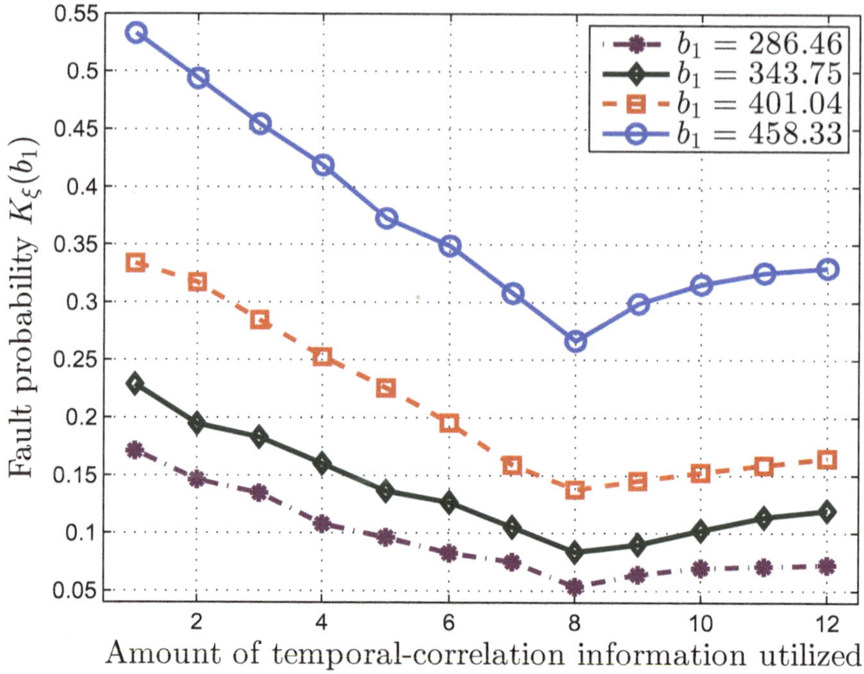

Fig. 5.5 Fault probability $K_\xi(b_1)$ with respect to the amount of temporal-correlation information utilized under different value of robust EA decision b_1

decision and plot Fig. 5.5. As depicted in this figure, when we expand the time lag window to utilize more temporal-correlation information, the fault probability will decrease first and then increase. This result indicates that the temporal-correlation information of solar energy generation within a proper time span is of benefit for reducing the conservativeness of the robust solution, whereas the correlation information outside a time lag (8 h) window is useless, even harmful for the decision-making. Thus, we may suggest that the microgrid should only utilize the temporal-correlation information within 8 h for developing the scheduling strategies.

5.6 Conclusion

In this chapter, we investigated the energy generation scheduling problem in a microgrid system equipped with renewable energy resources and energy storage devices. The aim of the scheduling is to minimize the system operation cost while maintaining the system reliability. To cope with the indeterminacy nature of renewable energy generation, we adopted a moment statistic model to confine the fluctuations. Such model allows convenient handling of volatile renewable energies as long as the generations are not too intensely different from the predictions or empirical knowledge. Chance-constraint approximations and robust optimization approaches based on generalized Chebyshev bounds are developed to first transform and then solve the scheduling problem. Numerical results based on real-world statistics evaluate the cost bounds of the proposed scheduling scheme. The impact of different parameters has been carefully studied. Moreover, we investigated the temporal-correlation properties of the solar energy. It is shown that the temporal-correlation information of solar energy generation within a proper time lag is beneficial for reducing the conservativeness of the robust solution, whereas the correlation information of longer time span may be harmful for the decision-making. These results, as we believe, shall provide useful insights helping the microgrid system operators to develop rational scheduling strategies.

References

1. D. Yang, C. Gu, Z. Dong, P. Jirutitijaroen, N. Chen, W.M. Walsh, Solar irradiance forecasting using spatial-temporal covariance structures and time-forward kriging. Renew. Energy **60**, 235–245 (2013)
2. T.K. Wijaya, S.F.R.J. Humeau, M. Vasirani, K. Aberer, Residential electricity load forecasting: evaluation of individual and aggregate forecasts. Technical report **EPFL** (2014)
3. R. Wang, P. Wang, G. Xiao, S. Gong, Power demand and supply management in microgrids with uncertainties of renewable energies. Int. J. Electr. Power Energy Syst. **63**, 260–269 (2014)
4. R. Wang, P. Wang, G. Xiao, A robust optimization approach for energy generation scheduling in microgrids. Energy Convers. Manag. **106**, 597–607 (2015)
5. X. Tan, Y. Wu, D.H. Tsang, A stochastic shortest path framework for quantifying the value and lifetime of battery energy storage under dynamic pricing. IEEE Trans. Smart Grid **8**, 769–778 (2017)
6. B.P.G. Van Parys, P.J. Goulart, D. Kuhn, Generalized gauss inequalities via semidefinite programming. Math. Program. **156**(1), 271–302 (2016)
7. D. Yang, P. Jirutitijaroen, W.M. Walsh, Hourly solar irradiance time series forecasting using cloud cover index. Sol. Energy **86**(12), 3531–3543 (2012)
8. California commercial end-use survey (2013)
9. Tecogen: advanced modular chp systems (2012)

Chapter 6
Massive Electric Vehicle Charging Involving Renewable Energy

6.1 Introduction

In the world today, fossil fuels are the dominant energy sources for both transportation sector and electricity generation industry. Statistics show that transportation and electricity generation account for over 60% of global primary energy demands [1]. The future solution for the fossil fuels scarcity, as well as the growing environmental problems associated with their wide usage, will most likely involve an extensive use of electric vehicles (EVs) and adopting renewable energy sources for electric energy production [2]. Under such cases, renewable energy supplied EV charging is becoming a popular approach for greener and more efficient energy usage. Since EVs have controllable charging rate, they can be considered as flexible loads in grid system which can benefit the grid system with demand response or load following. Accordingly, charging scheduling of EVs in the presence of renewable energy becomes a practical and important research problem.

A number of technical and regulatory issues, however, have to be resolved before renewable energy supplied EV charging becomes a commonplace. The arrival of EVs and their required energy amount may appear to be random, which increases the demand-side uncertainties. In addition, while renewable energy offers a cheaper and cleaner energy supply, it imposes great challenges to the stability and safety of the charging system because of its high inter-temporal variation and limited predictability. Therefore, the stochastic characteristics of both EVs and renewable energy sources should be carefully considered. Standby generators, back-up energy suppliers, or bulk energy storage systems may be necessary to alleviate the unbalancing issue caused by renewable energy fluctuation, which results in extra cost. In order to minimize the cost for obtaining extra energy and to increase energy efficiency, a flexible and efficient EV charging mechanism has to be properly designed to dynamically coordinate the renewable energy generation and energy demands of EVs.

In this chapter, we consider charging scheduling of a large number of EVs at a charging station which is equipped with renewable energy generation devices. The charging station can also obtain energy through controllable generators or buying

© Springer Nature Singapore Pte Ltd. 2018

R. Wang et al., *Intelligent Microgrid Management and EV Control Under Uncertainties in Smart Grid*, https://doi.org/10.1007/978-981-10-4250-8_6

energy from outside power grid. Stimulated by the fact that in practical scenario, EV arrival and renewable energy may not follow any determinate process yet obtaining some statistical information of future EVs' arrivals (departures) is possible, we propose a novel two-stage EV charging mechanism to reduce the cost and efficiently utilize renewable energy. Several uncertain quantities such as the arrival and departure time of the EVs, their charging requirements and available renewable energies are all taken into account. In addition, the mechanism allows more information of EV arrivals (departures) and renewable energy generation to be effectively incorporated into the charging mechanism when such information is available. The main contributions of this chapter can be briefly summarized as follows:

- A day-ahead cost minimization problem is formulated and solved based on the available prediction of future renewable energy generation and EVs' arrivals (departures) to determine the amount of energy generated or imported in a day-ahead manner.
- We propose a real-time EV charging and power regulation scheme based on the planned energy generation day-ahead to determine the charging rate of each vehicle and power output adjustments in a dynamic and flexible manner.
- We develop a fast charging rate compression (CRC) algorithm which significantly reduces the complexity of solving the real-time EV charging scheduling problem. The proposed algorithm supports real-time operations and enables the large-scale small-step scheduling more efficiently.
- We further extend our mechanism to be applicable to two practical scenarios: (1) the charging station needs to track a given load profile; and (2) the EVs only have discrete charging rates.

Simulation results indicate that our proposed two-stage EV charging mechanism can effectively reduce the system expenditure and peak-to-average ratio (PAR). Moreover, the proposed mechanism enhances the system fault tolerance against renewable energy uncertainties and the noises of real-time data. Note that the proposed charging scheme adopts a universal methodology which is not restricted to the specific data traces used in the paper: as long as the renewable energy generation data and EVs pattern data (including EVs battery level, desired charging amount, charging speed, and arrival/departure times) can be obtained, the proposed EV charging scheduling scheme can be implemented with virtually no change.

The remainder of this chapter is organized as follows: Sect. 6.2 introduces the problem formulation and two-stage decision-making process. In Sect. 6.3, we present the fast charging compression algorithm. The simulation results and discussions are presented in Sect. 6.4. An extension of the proposed charging mechanism is discussed in Sect. 6.5. Finally, we conclude this chapter in Sect. 6.6.

6.2 Two-Stage Decision-Making Model and Problem Formulation

6.2.1 Two-Stage Decision-Making Model

As shown in Fig. 6.1, we consider a charging park where an intelligent controller is responsible for the charging scheduling of a large number of EVs. To meet the EVs' energy demands, the intelligent controller (1) acquires electricity from either controllable energy plants (a dedicated power supply [2]) or central power grid; and (2) harvests the renewable energy from local solar panels or wind turbines. Considering the practice of energy acquisition from controllable generators or power grid and the limited predictability of renewable energy, we propose a two-stage model for decision making as shown in Fig. 6.2. Specifically, at the first stage, we divide time into discrete time slots with equal length.[1] The preliminary energy acquisition profile $\tilde{E}_c(h)$ and energy transfer factor $\alpha(h)$ are determined day-ahead before dispatch based on the estimated EV energy demand $\tilde{E}_v(h)$ and renewable energy generation $\tilde{E}_r(h)$, where $h \in \mathcal{H}$ is the time slot index and $\mathcal{H}$ is the set of time slots in day-ahead scale. Note that $\tilde{E}_v(h)$ is computed through the EVs' arriving and departing pattern predictions. On the other hand, the supply of renewable power $P_r(t)$ and EVs' real power demand $P_v(t)$ at time t can only be known in real time, which requires the real-time control to balance the power supply and demand at the second stage (real-time stage) if necessary. Hence during the real-time EV charging scheduling, we try to obtain the proper EVs' charging rates $V_i(t)$ and real-time power acquisition $P_c(t)$ given the real-time renewable power generation $P_r(t)$, EVs' real-time parking profiles and day-ahead dispatched acquired power $\tilde{P}_c(t)$ (determined in the first stage). Note that for the first stage, the decision making is done one time day-ahead. For the second stage, it is done more frequently in real time, i.e., as long as the renewable power generation or the parking states change, the EVs' charging decision coordinates accordingly. Table 6.1 lists the main notations to be used in the rest of this chapter.

6.2.2 Modeling System Uncertainties

It can be noticed that the intelligent charging operation involves several uncertain quantities including power available from the renewable energy system, the EVs' arrival and departure time, and their required charging amount. These quantities are crucial parameters for managing the energy generation and consumption of the system. Although these quantities are random, there are good reasons to expect that

[1] For the day-ahead energy generation scheduling, the length of one time slot usually varies between 5 and 30 min (as indicated in p. 149, Ref. [3]). Typically, a smaller slot duration enables the energy generation scheduling more flexible; meanwhile it to a certain level complicated the computation process. The specific suitable time slot length depends on the scale of the charging system and accuracy of the demand and load predictions.

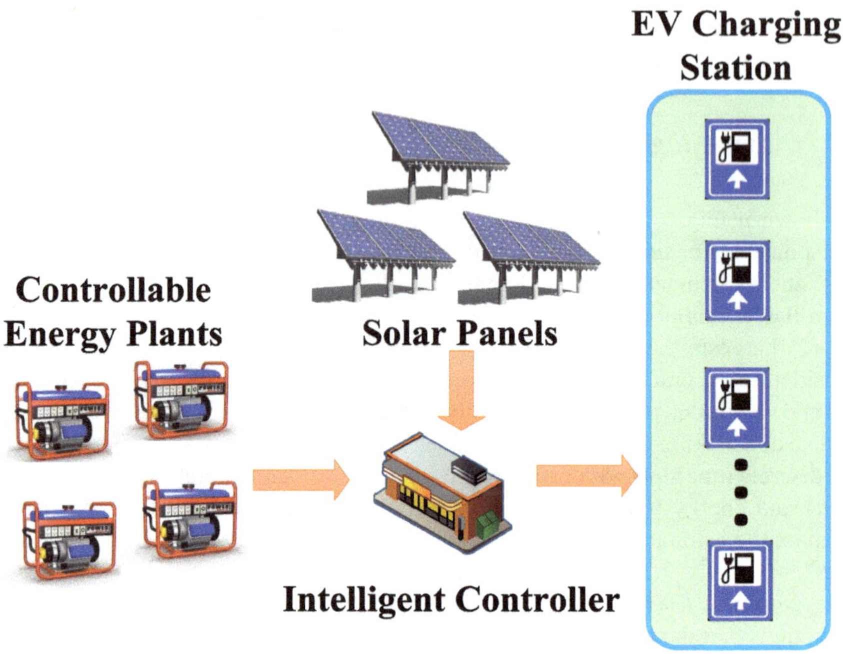

Fig. 6.1 The architecture of the EV charging station

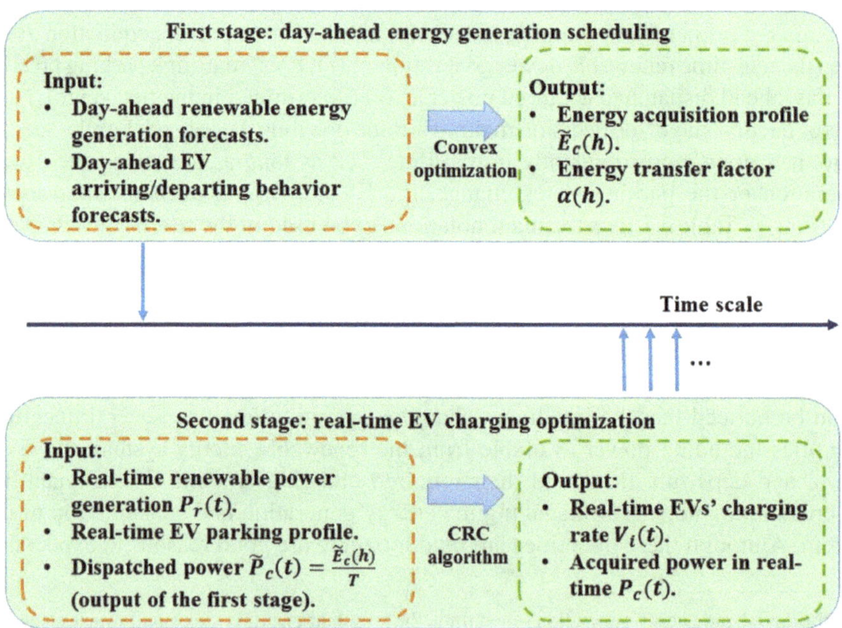

Fig. 6.2 Illustration of two-stage decision-making model. First stage (day-ahead): the decision variables are acquisition profile $\tilde{E}_c(h)$ and energy transfer factor $\alpha(h)$. Second stage (real-time): the decision variables are the charging speeds of EVs $V_i(t)$

Table 6.1 Notations used in this Chapter

Symbol	Definition
$\mathcal{H}$	Set of time slots in day-ahead scale, $\lvert\mathcal{H}\rvert = H$
h	Element in $\mathcal{H}$, time slot index in day-ahead scheduling
t	Time index in the real-time scheduling
$\tilde{E}_c(h)$	Predetermined energy acquisition at time slot h
$\tilde{E}_v(h)$	Estimated EV energy demand at time slot h
$\tilde{P}_v(t)$	Estimated EV power demand at time t
$\tilde{E}_r(h)$	Estimated renewable energy generation at time slot h
T	Length of one time slot
$\alpha(h)$	Energy transfer factor at time slot h
$M(t)$	The number of EVs in the charging park at time t
$w_i(t)$	Priority factor of EV i at time t
$V_{i_{max}}$	The maximum charging rate of EV i
$V_{i_{min}}$	The minimum charging rate of EV i
$V_i(t)$	Charging rate of vehicle i at time t
$V_d(t)$	The desired total charging demand at time t
$P_r(t)$	Renewable power realization at time t
$P_c(t)$	Power generated or imported in real-time
Γ	The set of charging tasks whose charging rates can vary
Γ_S	The set of charging tasks whose charging rates are fixed to maximum
τ_i	Charging task of EV i.

some statistical information may be obtained through accumulation of historical records. For example, the average energy generated by the renewable energy sources at each time slot can be estimated in a day-ahead manner based on the historical data and the weather forecast; inspecting a large number of samples of EVs' arrival and departure time, a probability distribution trend can be envisioned. We assume that the parking lot can roughly estimate the following parameters day-ahead: EVs arrival time distribution $f_A(x)$, departure time distribution $f_D(x)$, the total number of EVs being charged in a day $\bar{N}$, and the average charging rate of an EV μ_v. In this case, the estimated power (energy density) demand at time t can be expressed as:

$$\tilde{P}_v(t) = \int_0^t \big(f_A(x) - f_D(x)\big)\mathrm{d}x \cdot \bar{N} \cdot \mu_v, \tag{6.1}$$

and the estimated energy demand during time slot h is:

$$\tilde{E}_v(h) = \int_{h-1}^{h} \tilde{P}_v(t)\mathrm{d}t, \quad \forall h \in \mathcal{H}. \tag{6.2}$$

6.2.3 Day-Ahead Energy Acquisition Scheduling

The intelligent controller will firstly decide how much energy needs to be generated or imported in a day-ahead manner to minimize the expected energy acquisition cost while fulfilling the energy demand of EV charging station. The day-ahead energy acquisition scheduling problem can be formulated as:

$$\min_{\tilde{E}_c(h),\alpha(h)} \quad \sum_{h=1}^{H} \mathcal{C}_h\left(\tilde{E}_c(h)\right) \tag{6.3}$$

$$\text{s.t.} \quad \tilde{E}_c(h) + \tilde{E}_r(h) \geq \tilde{E}_v(h) \cdot \alpha(h) \tag{6.4}$$

$$\sum_{h=1}^{H} \tilde{E}_v(h) \cdot \alpha(h) = \sum_{h=1}^{H} \tilde{E}_v(h) \tag{6.5}$$

$$\alpha^L \leq \alpha(h) \leq \alpha^U, \forall h \in \mathcal{H}, \tag{6.6}$$

where $\mathcal{C}_h(\cdot)$ is the cost function of the electricity acquisition for the charging station, which is assumed to be an increasing convex function. The convex property reflects the fact that each additional unit of power needed to serve the demands is provided at a non-decreasing cost. Example cases include the quadratic cost function [4, 5] and the piecewise linear cost function [6, 7]. Without loss of generality, we consider quadratic cost function throughout this chapter. As to the renewable energy cost, for typical renewable energies (e.g., solar and wind energy), capital cost dominates. The operation and maintenance costs are typically very low or even negligible [8, 9]. In this chapter, it is assumed that the renewable energy generators such as solar panels and wind turbines have already been installed, and the marginal cost of renewable energy can be neglected, leading to its omission in the objective function [10]. Due to the flexibility of EVs' charging tasks, it is possible to shift some energy demand to other time slots to achieve the demand response target and reduce the total cost. $\alpha(h) > 0$ is an energy transfer factor, and $1 - \alpha(h)$ controls the portion of demand at time slot h shifted to other time slots. If $\alpha(h) > 1$, energy demand from other time slots is transferred to time slot h, whereas if $\alpha(h) < 1$, the energy demand in time slot h is shifted to other time slots. Note that $\alpha(h)$ can vary within its lower bound α^L and upper bound α^U. Constraint (6.4) is the load balance constraint, simply indicating that energy in each time slot should be balanced. Constraint (6.5) reveals the fact that the total energy required from EVs during a day remains unchanged, i.e., demand only transfers between time slots.

6.2.4 Real-Time Power Regulation and Elastic EV Charging

It is assumed that a two-way communication infrastructure (e.g., a local area network (LAN)) is available between the intelligent controller and vehicles. When an EV plugs in, it informs the intelligent controller its unplug time, desired charging amount, maximum and minimum allowable charging rates. Also, it is assumed that the EV owners are rational, so that the desired charging amount will not exceed the maximum charging capacity of vehicle during its parking period. In other words, if the vehicle is charged at its maximum speed during the entire parking period, it can definitely reach the preset desired battery level. For the real-time operation, the intelligent controller has two tasks. First, given the real renewable generation and EVs' charging requirements, it has to determine a proper charging rate for each EV to achieve the optimal utilization of renewable energy and finish the charging tasks before EVs' departures. Second, the total acquired power should be properly regulated around the predetermined generation profile in real-time to match the fluctuant power demand, i.e., demand and supply should be balanced at any time instance.

From the standpoint of EV owners, it is desirable to reduce their EVs' charging time. For example, decreasing the charging time provides more flexibility for the owners to leave the charging station earlier. This objective can be captured by the constrained optimization problem as follows:

$$\min_{V_i(t)} \sum_{\tau_i \in \Gamma} w_i(t)\big(V_{i_{max}} - V_i(t)\big)^2, \tag{6.7}$$

$$\text{s.t.} \sum_{\tau_i \in \Gamma} V_i(t) + \sum_{\tau_i \in \Gamma_S} V_{i_{max}} \le V_d(t), \tag{6.8}$$

$$V_i(t) \ge V_{i_{min}} \qquad \forall \tau_i \in \Gamma, \tag{6.9}$$

$$V_i(t) \le V_{i_{max}} \qquad \forall \tau_i \in \Gamma. \tag{6.10}$$

In (6.7), decision variable is $V_i(t)$ which is the charging speed of EV i to be determined at time t. τ_i represents the charging task of vehicle i. Parameter $w_i(t) \ge 0$ is a priority factor which reflects the urgent degree of a charging task. More urgent tasks would have larger $w_i(t)$. Without loss of generality, $w_i(t)$ can be determined dynamically according to the state of the EV, which is defined as follows:

$$w_i(t) = \frac{E_i^r}{T_i^d - t}, \quad \forall \tau_i \in \Gamma, \tag{6.11}$$

where E_i^r is the amount of remaining requested energy for charging and T_i^d is EV i's departure time. Equation (6.11) indicates that urgent charging tasks will have a higher priority factor so as to be charged faster. This is to ensure that EVs depart with desired battery level. w_i also denotes the average charging rate EV i needs to finish the charging task τ_i on time. $V_{i_{max}}$ is the maximum charging rate (i.e., the desired charging rate) of EV i. $V_{i_{min}}$ is the minimum allowable charging rate of EV i. At any time t, the charging tasks can be first classified into two categories: Γ is the set of charging tasks whose charging rate can vary, i.e., $\Gamma = \{\tau_i |\ w_i(t) < V_{i_{max}}\}$. Γ_S denotes the set of charging tasks whose charging rates have to be fixed at the maximum charging rates because of the urgent charging time, i.e., $\Gamma_S = \{\tau_i |\ w_i(t) = V_{i_{max}}\}$. Note that elements in Γ and Γ_S may vary with time and for $\tau_i \in \Gamma$, $V_i \le V_{i_{max}}$, for $\tau_i \in \Gamma_S$,

$V_i = V_{i_{max}}$. This EV classification approach ensures that all the EVs depart with satisfactory charging amount. $V_d(t)$ is the desired total charging demand at time t. The way to set $V_d(t)$ will be introduced later.

Notice that constraint (6.8) simply states the schedulability condition, and the rest of the constraints bound the charging rates. Due to EVs' arrivals and departures, the system is dynamic and the number of vehicles and their charging requirements will change over time. Therefore, the intelligent controller can solve problem (6.7)–(6.10) to obtain the charging rate for each EV at time t. When the renewable power realization changes, or an EV's status changes (τ_i changes from Γ to Γ_S) or a vehicle enters or departs the system, the intelligent controller will update Γ, Γ_S, and $V_d(t)$ in real time and then redo the calculation. Next, we will show how to determine $V_d(t)$ to optimally utilize the renewable energy.

Let $\tilde{P}_c(t) = \frac{\tilde{E}_c(h)}{T}$ denote the dispatched acquired power (i.e., the day-ahead pre-scheduled power generation) at time t, where T is the length of a time slot, and $P_r(t)$ denote the renewable generation realization at time t. Then, $V_d(t)$ can be defined as follows:

- If $\sum_{\tau_i \in \Gamma} V_{i_{min}} + \sum_{\tau_i \in \Gamma_S} V_{i_{max}} > \tilde{P}_c(t) + P_r(t)$, then $V_d(t) = \sum_{\tau_i \in \Gamma} V_{i_{min}} + \sum_{\tau_i \in \Gamma_S} V_{i_{max}}$, $P_c(t) = V_d(t) - P_r(t)$, $P_c(t)$ is the acquired power in real time. This is for the case where the renewable energy generation is very low, i.e., even though all the controllable EVs (EVs that belong to set Γ) charge at their minimum allowable charging rates, the demand is still higher than the available supply. Therefore, up regulation is required to guarantee the power balancing, i.e., more energy has to be imported, either by raising up the output level of fast-response generators or buying more electricity from ancillary service markets.
- If $\sum_{\tau_i \in \Gamma} V_{i_{min}} + \sum_{\tau_i \in \Gamma_S} V_{i_{max}} \leq \tilde{P}_c(t) + P_r(t) \leq \sum_{\tau_i \in \Gamma \cup \Gamma_S} V_{i_{max}}$, then $V_d(t) = \tilde{P}_c(t) + P_r(t)$ and $P_c(t) = \tilde{P}_c(t)$. This investigates the scenario where the renewable energy generation deviates not far from the previous prediction, i.e., the power demand of EVs can be adjusted to match the available supply. This represents the most common situation the charging system encounters. Under such case, the power demand of controllable EVs can be adjusted to match the supply, thus power acquisition profile does not need to be changed and is equal to the dispatched load determined day-ahead.
- If $\sum_{\tau_i \in \Gamma \cup \Gamma_S} V_{i_{max}} < P_r(t) + \tilde{P}_c(t)$, then $V_d(t) = \sum_{\tau_i \in \Gamma \cup \Gamma_S} V_{i_{max}}$ and $P_c(t) = \sum_{\tau_i \in \Gamma \cup \Gamma_S} V_{i_{max}} - P_r(t)$. This corresponds to the case where the renewable energy generation is plenty enough that even the highest charging demand can be satisfied, i.e., although all the EVs charge at the maximum charging rates, available power still exceeds. In this case, down regulation is required to make sure that power is balanced, i.e., the intelligent controller can reduce the acquired power level or sell the extra power out and only compensate the mismatch between the maximum charging demand and the renewable energy output.

Remark In day-ahead energy acquisition scheduling, the intelligent controller aims at minimizing the expected cost of the charging park given the estimated renewable energy supply $\tilde{E}_r(h)$ and EVs' energy demand $\tilde{E}_v(h)$, $h \in \mathcal{H}$. Decision variable

$\tilde{E}_c(h)$ is the scheduled electricity to be brought from day-ahead energy market or generated by base-load plants. In real-time power regulation, system reliability and EVs' charging requirements become the main concerns. The aforementioned up/down regulation is provided by ancillary service markets or fast-response generators [11].

6.3 The Charging Rate Compression Algorithm

The problems (6.7)–(6.10) belongs to the category of convex quadratic programs and can be solved in polynomial time. Many commercial optimization solvers including CPLEX, Mosek, FortMP, and Gurobi can be utilized to solve such problems. However, solving such a problem using quadratic program solver during run time can be still too costly, especially when the number of EVs is large and the response time has to be very short so as to quickly respond to EVs. What makes the above formulation attractive is that a charging rate compression (CRC) algorithm can be proposed such that the problem solving can be extremely fast. We first develop the CRC algorithm and then introduce a lemma and a theorem to prove that it can solve the problems (6.7)–(6.10).

At each time instance t, the set Γ of charging tasks can be further divided into two subsets: a set Γ_f of charging tasks with the minimum charging rate and a set Γ_v of charging tasks whose charging rate can still be compressed. Let $V_0 = \sum_{i\in\Gamma} V_{i_{max}}$ be the maximum power level of the charging task set Γ, V_{v0} be the sum of maximum charging rates of charging tasks in Γ_v, and V_f be the sum of the charging rates of charging tasks in Γ_f. To achieve a desired power level $V_d(t) < V_0 + \sum_{i\in\Gamma_S} V_{i_{max}}$, each charging task has to be compressed up to the following charging rate:

$$\forall \tau_i \in \Gamma_v, \quad V_i = V_{i_{max}} - (V_{v0} - V_m(t) + V_f)\frac{W_v}{w_i}, \tag{6.12}$$

where

$$V_m(t) = V_d(t) - \sum_{\tau_i \in \Gamma_S} V_{i_{max}} \tag{6.13}$$

$$V_{v0} = \sum_{\tau_i \in \Gamma_v} V_{i_{max}} \tag{6.14}$$

$$V_f = \sum_{\tau_i \in \Gamma_f} V_{i_{min}} \tag{6.15}$$

$$W_v = \frac{1}{\sum_{\tau \in \Gamma_v} \frac{1}{w_i}}. \tag{6.16}$$

If there exist charging tasks where $V_i < V_{i_{min}}$, then the charging rates of these vehicles have to be fixed at their minimum value $V_{i_{min}}$. Sets Γ_f and Γ_v have to be updated (therefore, V_f, V_{v0}, and W_v have to be recomputed), and (6.12) is applied again to the charging tasks in Γ_v. If a feasible solution exists, i.e., the desired power level of the system is higher than or equal to the minimum power level $\sum_{i=1}^{M(t)} V_{i_{min}}$, the iterative process ends until each value computed by (6.12) is greater than or equal to its corresponding minimum $V_{i_{min}}$. The algorithm for compressing the charging rate of a set Γ of EVs to a desired charging power level $V_d(t)$ is shown in **Algorithm 6.1**.

Algorithm 6.1 Algorithm for compressing the charging rate for a charging task set of Γ at time t.

Input: $V_d(t)$, $V_{i_{min}}$, $V_{i_{max}}$, w_i, $\forall \tau_i \in \Gamma$.
Output: V_i, $\forall \tau_i \in \Gamma$.

Begin
$V_0 = \sum_{\tau_i \in \Gamma} V_{i_{max}}$;
$V_{min} = \sum_{\tau_i \in \Gamma} V_{i_{min}}$;
$V_m(t) = V_d(t) - \sum_{\tau_i \in \Gamma_S} V_{i_{max}}$;
if $(V_m(t) < V_{min})$
  Return INFEASIBLE;
else
  do {
    $\Gamma_f = \{\tau_i | V_i = V_{i_{min}}\}$;
    $\Gamma_v = \Gamma - \Gamma_f$;
    $V_{v0} = \sum_{\tau_i \in \Gamma_v} V_{i_{max}}$;
    $V_f = \sum_{\tau_i \in \Gamma_f} V_{i_{min}}$;
    $W_v = \frac{1}{\sum_{\tau \in \Gamma_v} \frac{1}{w_i}}$;
    OK= 1;
    for (each $\tau_i \in \Gamma_v$)
      $V_i = V_{i_{max}} - (V_{v0} - V_m(t) + V_f)\frac{W_v}{w_i}$;
      if $(V_i < V_{i_{min}})$
        $V_i = V_{i_{min}}$;
        OK= 0;
      end **if**
    end **for**
  } **while** (OK== 0);
  return FEASIBLE;
end **if**
End

Lemma 6.1 *Given the constraint optimization problem as specified in (6.7)–(6.10) and* $\sum_{\tau_i \in \Gamma} V_{i_{max}} > V_m(t)$*, any solution,* $V_i^*(t)$*, to the problem must satisfy* $\sum_{\tau_i \in \Gamma} V_i^*(t) = V_m(t)$ *and* $V_i^*(t) \neq V_{i_{max}}$*, for all* $\tau_i \in \Gamma$.

Theorem 6.1 *Given the constraint optimization problem as specified in (6.7)–(6.10),* $\sum_{\tau_i \in \Gamma} V_{i_{max}} > V_m(t)$*, and* $\sum_{\tau_i \in \Gamma} V_{i_{min}} < V_m(t)$*, let* $\widehat{V}(t) = \sum_{V_i^*(t) \neq V_{i_{min}}} V_{i_{max}} + \sum_{V_i^*(t) = V_{i_{min}}} V_{i_{min}}$*. A solution is optimal if and only if*

$$V_i^*(t) = V_{i_{max}} - \frac{\frac{1}{w_i(t)}(\widehat{V}(t) - V_m(t))}{\sum_{V_j^*(t) \neq V_{j_{min}}} (1/w_j)}, \tag{6.17}$$

for $\widehat{V}(t) > V_m(t)$ *and* $V_i^*(t) > V_{i_{min}}$, *and* $V_i^*(t) = V_{i_{min}}$ *otherwise.*

The proofs of Lemma 6.1 and Theorem 6.1 are given in the **Appendix B**. Based on the previous lemma and theorem, we can draw the conclusion as follows:

Corollary 6.1 *Consider the charging tasks of* $|\Gamma \cup \Gamma_S|$ *EVs, where* $V_i(t)$ *is the charging rate of the* i*th vehicle. Let* $V_{i_{max}}$ *denote the initial desired charging rate of charging task* $\tau_i \in \Gamma \cup \Gamma_S$ *and* $w_i(t)$ *be the set of priority factors. Let* $V_d(t)$ *be the desired power level of the system and* $\sum_{\tau_i \in \Gamma} V_{i_{max}} > V_m(t)$. *The charging rate* V_i, $\tau_i \in \Gamma$, *obtained from* ***Algorithm 6.1*** *minimizes*

$$\sum_{\tau_i \in \Gamma} w_i(t)\left(V_{i_{max}} - V_i(t)\right)^2$$

subject to the inequality constraints $\sum_{\tau_i \in \Gamma} V_i(t) + \sum_{\tau_i \in \Gamma_S} V_{i_{max}} \le V_d(t)$, $V_i(t) \ge V_{i_{min}}$, *and* $V_i(t) \le V_{i_{max}}$ *for* $\tau_i \in \Gamma$.

Remark Through analysis, the time complexity of **Algorithm 6.1** is $O(n^2)$, where n is the number of tasks in Γ.

6.4 Simulation Results and Discussions

In this section, we present simulation results based on real-world traces for assessing the performance of the proposed two-stage EV charging scheme.

6.4.1 Parameters and Settings

We assume there are solar panels providing renewable energy for the charging station. The area of the solar panels in the system is set to be $3.125 \times 10^4\,\text{m}^2$. The energy conversion efficiency is 0.8. The solar radiation intensity statistic is adopted from [12], from which we employ the solar radiation data of a typical day in winter (17/01/2013). The data utilized for the day-ahead energy acquisition scheduling and real-time EV charging are depicted in Fig. 6.3. Note that the predicted average solar radiance utilized in the day-ahead energy generation scheduling is plotted in the blue circled line, and the actual real-time solar radiance adopted in the real-time charging is shown by the red curve. We envision the scenario that the charging station is located at a workplace (e.g., a campus) that is active from 6:00 AM to 6:00 PM. Vehicles arrive earlier than 6:00 AM start to charge at 6:00 AM while those depart later than 6:00 PM finish their charging before 6:00 PM. We simulate the operation process of a large-scale charging station which serves totally 3000 EVs arriving and departing independently in a typical day. It is assumed that the arrival time distribution and departure time distribution are all Gaussian with parameters shown in Table 6.2

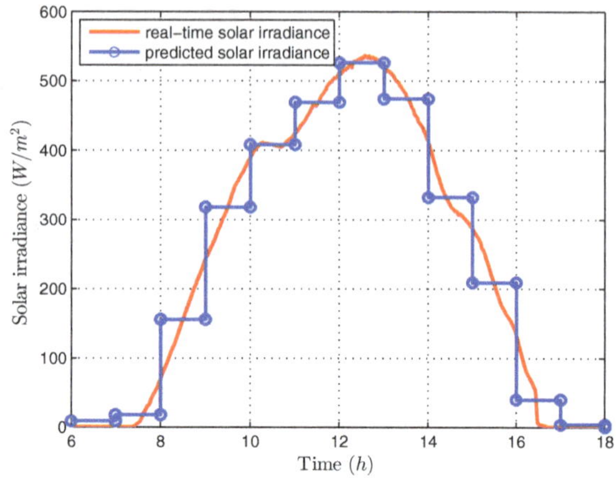

Fig. 6.3 Solar irradiance in a day

Table 6.2 Parameters of the arrival and departure time probability distribution

Time parameter	Arrival	Departure
Mean: μ_r	10	14
Standard deviation: σ_r	1.2	1.3

(similar assumptions can be found in many papers, e.g., [13, 14]). EVs are active for charging during their parking time, and discharging is not permitted. The amount of energy needed for the EVs are evenly distributed between 20 KWh and 50 KWh. The maximum allowable charging rate of an EV is 62.5 KW (e.g., high-voltage (up to 500 VDC) high-current (125 A) automotive fast charging [15]), and the minimum charging rate of an EV is 0 KW. The cost function of the electricity acquisition is $C_h\left(\tilde{E}_c(h)\right) = a_h \cdot \tilde{E}_c(h)^2$ and $a_h = 150\ \$ \cdot (\text{MWh})^{-2}$.

6.4.2 *Results and Discussions*

The simulation process contains two parts. First, given the estimated solar energy in each time slot (in the simulation, one time slot is set as one hour), we solve the day-ahead energy acquisition scheduling problems (6.3)–(6.6) and obtain $\tilde{E}_c(h)$ and $\alpha(h)$ for $h = 1, \ldots, H$. The upper bound and lower bound of energy transfer parameter $\alpha(h)$ is set to be 2 and 0.5, respectively. Once the dispatched energy acquisition in each time slot is obtained, we are ready to simulate the charging process of EVs based on the real-time renewable power generation and EVs' real-time arrival (departure) patterns. Adopting the data previously mentioned, all the simulations are conducted on an Intel workstation with six processors clocking at 3.2 GHZ and 16 GB of RAM. We repeated the simulation for 10 times. All the 3000

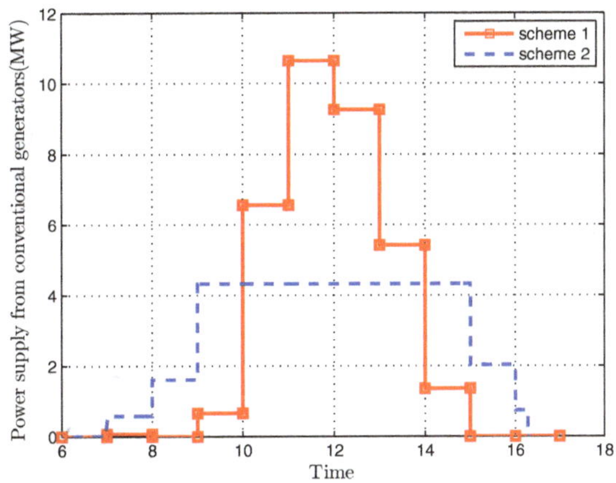

Fig. 6.4 Energy supply from conventional generators under different charging schemes

EVs complete charging with required amount before their departures. By utilizing the CRC algorithm introduced in Sect. 6.3, the simulation time is reduced from 1005.1 s to 101.2 s, showing that the proposed CRC algorithm can significantly reduce the complexity of the problem solving. Note that our CRC algorithm does not sacrifice the problem-solving accuracy, and we obtain exactly the same results when adopting quadratic programming solvers and our CRC algorithm.

We first investigate the effectiveness of our proposed EV charging mechanism. Specifically, two charging schemes are compared. In the first scheme, EVs are kept charging during their parking time and the charge speeds are the average rates that they need to fulfill the charging tasks. Conventional generators generate electricity for the unbalanced power demand in an on-demand manner. While in the second scheme, the charging station charges EVs' batteries according to the mechanism we proposed, and electricity is generated based on the day-ahead scheduling and real-time adjustment. The simulation results concerning the power supply curves, total system cost, and peak-to-average ratio (PAR) under these two schemes are given in Figs. 6.4 and 6.5, respectively. As we mentioned previously, quadratic cost functions are adopted to compute the system expenditures for both schemes.

In Fig. 6.4, it is shown that by optimally controlling the charging rates of EVs, our proposed charging strategy successfully transfers the peak demand to the off-peak hours, which can help stabilize the operations of the charging system and reduce the energy cost. As shown in Fig. 6.5, the total expenditure of the charging station decreases from \$4.1 $\times$ 10^4 per day in scheme 1 to \$1.8 $\times$ 10^4 per day in our proposed scheme, achieving a cost saving of 56.1%. Therefore, one of the aims of the developed charging strategy, which is reducing the expenditure of the system, is achieved. To investigate the variation of PAR, we study two cases: (1) PAR of the aggregated supply (i.e., the supply from controllable generators plus the supply from solar panels); and (2) PAR of the controllable generators' output. As we observe in Fig. 6.5, with scheme 1, the PAR of the aggregated supply and the PAR of controllable generators' output are 2.69 and 3.95, respectively. By adopting the proposed charging

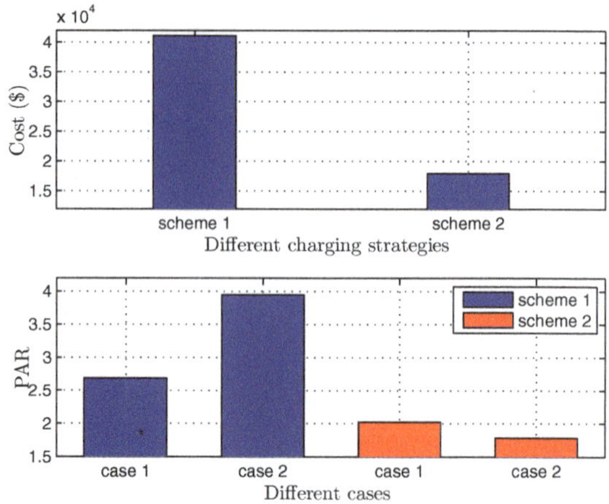

Fig. 6.5 Cost and PAR comparisons of different charging schemes

scheme, these two PAR values reduce to 2.02 and 1.78 (decrease about 25 and 55%), respectively. The proposed EV charging strategy presents much better PAR performance during the 12-h operation. An interesting observation is that in scheme 1, the PAR of controllable generators' output is much higher than that of the aggregated power supply; however, the situation is exactly opposite in our proposed scheme. In other words, under normal circumstances, utilizing renewable energy will make the output of controllable generators more fluctuant, whereas EVs can help solve this problem by properly varying their charging speeds, i.e., charging quickly when renewable energy is sufficient and reducing the rate when not enough renewable energy is available.

In our scheme, the first-stage day-ahead energy generation scheduling is based on the estimated renewable energy generation in next day. Normally, the real renewable energy generation might be different from the estimated one. Next, we investigate the cost sensitivity with respect to this deviation. The simulation results are depicted in Fig. 6.6. Specifically, we conduct the experiment as follows. In the first step, the day-ahead energy generation scheduling is done based on the estimated solar irradiance and EVs' arriving (departing) patterns. Then, for the real-time charging, we vary the solar irradiance data based on the real-world trace to represent different estimation error levels. As it is observed in Fig. 6.6, system cost is much more sensitive in scheme 1 than that in our scheme when the deviation varies. The reason is that by applying our charging strategy, deviations of the solar power can be distributed to the whole time horizon. However in scheme 1, the situation that solar power is excessive during some time periods and insufficient in some other time becomes more severe. Under such case, solar energy utilization efficiency fluctuates more extensively when deviation level increases, and accordingly, system cost varies more violently. Hence, our charging mechanism can effectively reduce the financial risks caused by the estimation error of the renewable energy generation.

Figure 6.7 illustrates how system cost varies under different fluctuation levels of solar energy. In this experiment, we add 0-mean Gaussian noise to the real-time solar

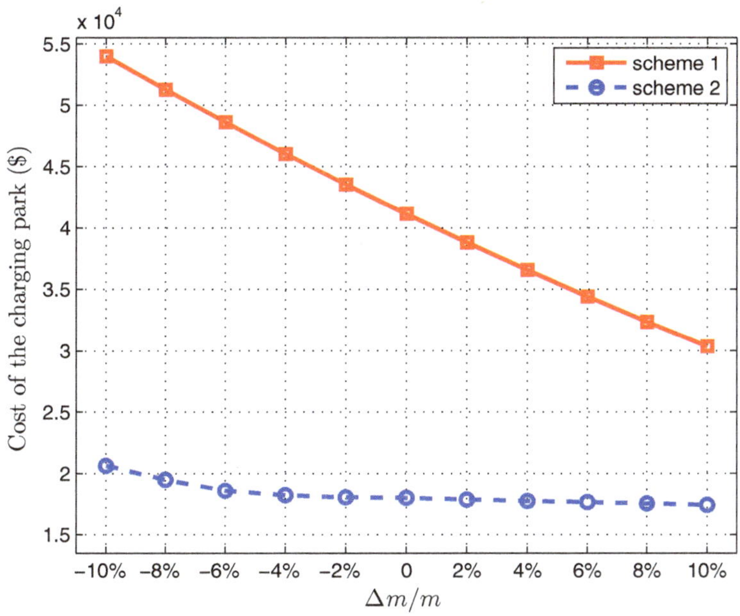

Fig. 6.6 System cost with respect to the real-time renewable generation deviation (Δm represents the deviation of real solar irradiance from the estimated one, and m is the actual data trace)

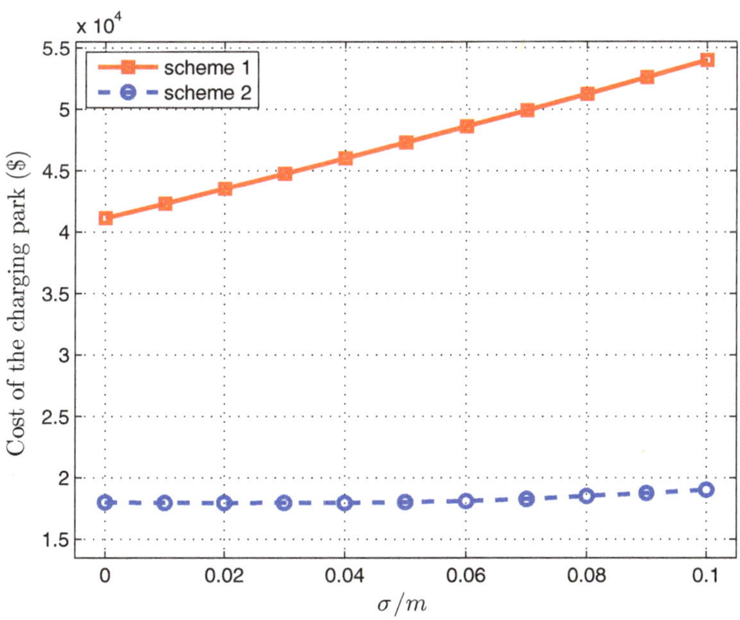

Fig. 6.7 System cost with respect to the different fluctuation level of renewable energy (m represents the actual data trace, and σ represents the standard deviation of noise)

irradiance data and then evaluate its impact on the system cost. Different standard deviations of the noise reflect different fluctuation levels of solar energy. It appears that the fluctuation of renewable energy has less impact on the system cost when adopting our proposed scheduling scheme. This observation is intuitive since by properly altering their charging rates, EVs act as an energy storage which may to a certain extent alleviate the uncertainty problem. However in scheme 1, the controllable generators have to compensate the solar power fluctuation during the entire time horizon. In this case, the system cost will be affected more extensively when fluctuation level increases. Note that this experiment also simulates the scenario that system data is affected by noises. Thus, we claim that the our proposed EV charging mechanism shows good performances in dealing with uncertainties of renewable energy and noises of real-time data.

6.5 Extensions

6.5.1 Tracking a Given Load Profile

The electricity utilized for EV charging can be provided by a utility company. The objective of the utility company may be to flatten the total load profile. The utility company may also need to buy electricity in day-ahead electricity market and supply the electricity to the charging parking as well as other energy consumers in real-time. Under such case, the utility company may want the charging station to properly schedule the charging of EVs, so that the demand can track the electricity profile it brought in the day-ahead electricity market. Denote the load profile that the charging park tracks as $L(t)$. Our charging scheme can be extended to track $L(t)$ by solving the following constraint optimization problem:

$$\min_{V_i(t)} \sum_{\tau_i \in \Gamma} w_i(t)\big(V_{i_{max}} - V_i(t)\big)^2, \tag{6.18}$$

$$\text{s.t.} \sum_{\tau_i \in \Gamma} V_i(t) + \sum_{\tau_i \in \Gamma_S} V_{i_{max}} \leq L(t), \tag{6.19}$$

$$V_i(t) \geq V_{i_{min}} \qquad \forall \tau_i \in \Gamma, \tag{6.20}$$

$$V_i(t) \leq V_{i_{max}} \qquad \forall \tau_i \in \Gamma. \tag{6.21}$$

Figure 6.8 shows the simulation results of tracking given target load profiles. The intelligent controller is in charge of managing 3000 EVs in a day on their charging schedules. These vehicles plug in uniformly distributed between 6:00 and 14:00, with deadlines uniformly distributed between 10:00 and 18:00. The amount of energies needed to charge are evenly distributed between 20 KWh and 50 KWh. Two testings are conducted to show the load tracking results with different target profiles. The target profiles are represented by the blue dot-circled curves. The red

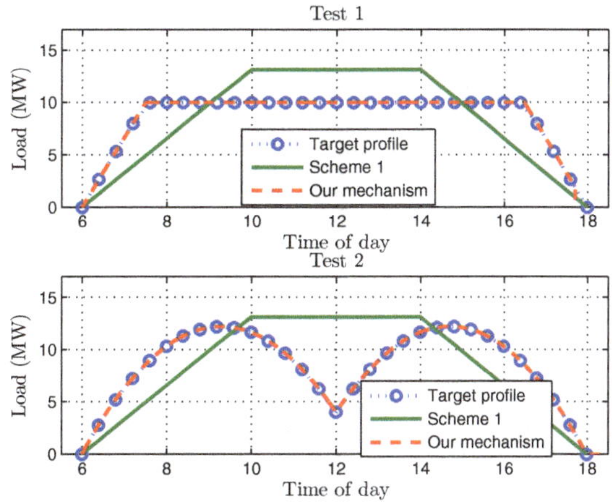

Fig. 6.8 Tracking given target load profiles

dash curves and green solid curves correspond to the aggregated charging rates obtained from our EV charging mechanism and scheme 1, respectively. We observe that the aggregated charging demand can closely follow the target load profiles when adopting our proposed charging scheme. There are only small discrepancies around 18:00 due to the early or late departures of EVs.

Remark In order to ensure that the electricity demand of EVs can closely follow the target load profile, load profile $L(t)$ should not go beyond the variation limits of EVs' charging rates, that is:

$$L(t) \geq \sum_{\tau_i \in \Gamma} V_{i_{min}} + \sum_{\tau_i \in \Gamma_S} V_{i_{max}} \tag{6.22}$$

and

$$L(t) \leq \sum_{\tau_i \in \Gamma} V_{i_{max}} + \sum_{\tau_i \in \Gamma_S} V_{i_{max}} \tag{6.23}$$

6.5.2 Discrete Charging Rates

In our proposed charging scheme, we assume that the charging rate can vary continuously within the EV's maximum and minimum allowable rates, determined by the charger. Similar assumptions can be found in many literature including [16–18]. However in some circumstances, if only a few discrete charging speeds are allowed, the proposed EV charging scheme can be easily extended to handle such case. Let $\mathcal{V}_i$ denote the set of allowable charging rates of vehicle i. To capture the discrete charging rate case, we replace constraints (6.9) and (6.10) with the following constraint in

Table 6.3 Simulation results under continuous charging rate case and discrete charging rate case (all results are 10 times average)

Charging park size	Power level type	Charging cost ($)	Cost growth (%)
Large scale (3000 EVs)	Continuous charging rate case	17993.1	–
	Discrete charging rate case	18028.8	0.2
Medium scale (500 EVs)	Continuous charging rate case	497.1	–
	Discrete charging rate case	511.2	2.8
Small scale (100 EVs)	Continuous charging rate case	22.2	–
	Discrete charging rate case	27.9	25.7

real-time EV charging:

$$V_i(t) \in \mathcal{V}_i \qquad \forall \mathcal{T}_i \in \Gamma. \tag{6.24}$$

Although the CRC algorithm is only suitable for the continuous charging rate case, simulations show that with discrete allowable charging rates, the proposed two-stage charging mechanism still has an acceptable computation-time performance. In the simulation, each EV has four allowable charging speeds, i.e., $\mathcal{V}_i = \{0\text{ KW}, 20\text{ KW}, 40\text{ KW}, \text{ and } 62.5\text{ KW}\}, \forall \mathcal{T}_i \in \Gamma$ [2]. The number of EVs served in a day is still 3000. The simulation results comparison with the continuous charging rate case is summarized in the first row of Table 6.3. Note that the simulations under both cases are conducted 10 times, and results in Table 6.3 are the average.

As it is shown in Table 6.3, two main observations can be found as follows:

- For the discrete charging rate case, though the simulation time is much longer for the continuous charging rate case, our two-stage EV charging mechanism still performs acceptably for the real-time scheduling since computation time for updating the charging rates of active vehicles is about 0.25 s on average. Note that this is the updating time running on the computer whose configuration is specified in the previous subsection.
- The system cost increases slightly (about 0.2%) when only several discrete charging rates are allowed. This observation is intuitive since with discrete charging rates, the scheduling flexibility is abated and mechanism performance gets worse. In other words, when EVs' charging rates can vary continuously, the power demand can follow the desired power supply more closely and thus utilize the renewable energy in a more efficient manner. However, since the number of EVs is large, discrepancy between the combinations of EVs' discrete charging rates and desired energy supply level is not significant. Thus, the cost only increases slightly.

We further reduce the simulation scale to medium size (e.g., 500 vehicles) and small size (e.g., 100 vehicles) to investigate how the size of the charging park impacts the performances of the proposed scheme. Besides the charging park size, simulation process and system parameters are exactly the same to those in the previous subsection. The area of the solar panels varies proportionally with the charging park size. We also simulate 10 times and the results' data are depicted in Table 6.3. It appears that for large-, medium-, and small-scale charging parks, system costs in discrete charging rate case are 0.2, 2.8, and 25.7% higher than those in the continuous charging rate case, respectively. In other words, system cost is more sensitive to the discrete charging rate condition when the scale of charging park shrinks. The reason for this phenomenon is that when the number of connected vehicles gets small and only several discrete charging rates are allowed, the flexibility of the system deteriorates. There will be a higher probability that aggregated charging demand cannot match the available power. For instance, when there are only 30 KW power available and two vehicles are active at a given time, for the continuous charging rate case, EVs are able to follow the supply closely. Whereas for the discrete charging rate case, either 10 KW power is wasted or conventional units have to generate 10 KW more so that discrete demand can be matched. Therefore, power utilization becomes less efficient, and conventional generators have to produce more electricity to ensure that charging tasks can be finished in time. As we mentioned previously, when the number of EVs is large, discrepancy between the combinations of EVs' discrete charging rates and desired energy supply level becomes less significant, leading to only marginal increase in cost. The proposed EV charging scheme favors reasonably for a large charging park when only discrete charging rates are allowed.

6.6 Conclusion

In this chapter, we investigate the cost-effective scheduling approach of EV charging at a renewable energy aided charging station. We design a two-stage EV charging scheme to determine energy generation and charging rates of EVs. Specifically, at the first stage, based on the EV pattern and renewable energy generation estimation, a cost minimization problem is formulated and solved to obtain a preliminary energy generation or importation scheduling profile in a day-ahead manner. Then at the second stage, a real-time EV charging and power regulation scheme are proposed. Such a scheme allows convenient handling of volatile renewable energy and indeterminate EV patterns. We also develop an efficient charging compression algorithm to further lower the complexity of the problem solving. Simulation results indicate the satisfactory efficiency of the proposed EV charging mechanism and the cost benefits obtained from it. Moreover, the impacts of renewable energy uncertainties have been carefully evaluated. The results show that the proposed EV charging scheme has a good performance in enhancing the system fault tolerance against uncertainties and the noises of real-time data. Such evaluations, as we believe, reveal that the proposed charging mechanism is suitable for the case with a large number of EVs and unstable renewable energy. Furthermore, we extend the mechanism to track a given load profile and handle the scenario that EVs only have discrete charging rates. As a universal methodology, the proposed scheme is not restricted to any specific data traces and can be easily applied to many other cases as well.

References

1. D.B. Richardson, Electric vehicles and the electric grid: a review of modeling approaches, impacts, and renewable energy integration. Renew. Sustain. Energy Rev. **19**, 247–254 (2013)
2. F. Mwasilu, J.J. Justo, E.-K. Kim, T.D. Do, J.-W. Jung, Electric vehicles and smart grid interaction: a review on vehicle to grid and renewable energy sources integration. Renew. Sustain. Energy Rev. **34**, 501–516 (2014)
3. C. Harris, *Electricity Markets: Pricing, Structures and Economics* (Wiley, New York, 2006)
4. A. Mohsenian-Rad, V. Wong, J. Jatskevich, R. Schober, A. Leon-Garcia, Autonomous demand-side management based on game-theoretic energy consumption scheduling for the future smart grid. IEEE Trans. Smart Grid **1**(3), 320–331 (2010)
5. P. Samadi, H. Mohsenian-Rad, R. Schober, V. Wong, Advanced demand side management for the future smart grid using mechanism design. IEEE Trans. Smart Grid **3**(3), 1170–1180 (2012)
6. C. Joe-Wong, S. Sen, S. Ha, M. Chiang, Optimized day-ahead pricing for smart grids with device-specific scheduling flexibility. IEEE J. Sel. Areas Commun. **30**(6), 1075–1085 (2012)
7. I. Koutsopoulos, L. Tassiulas, Optimal control policies for power demand scheduling in the smart grid. IEEE J. Sel. Areas Commun. **30**(6), 1049–1060 (2012)
8. Renewable power generation costs in 2012: an overview (2012)
9. Handbook for solar photovoltaic (pv) systems (2011)
10. Y. Guo, M. Pan, Y. Fang, Optimal power management of residential customers in the smart grid. IEEE Trans. Parallel Distrib. Syst. **23**(9), 1593–1606 (2012)
11. B. Kirby, Ancillary services: Technical and commercial insights. Retrieved October, vol. 4, p. 2012 (2007)
12. Nrel: National renewable energy laboratory (2013)
13. C. Jin, X. Sheng, P. Ghosh, Energy efficient algorithms for electric vehicle charging with intermittent renewable energy sources. In *IEEE Power and Energy Soc. General Meeting*, pp. 1–5, IEEE (2013)
14. A. Mohamed, V. Salehi, T. Ma, O. Mohammed, Real-time energy management algorithm for plug-in hybrid electric vehicle charging parks involving sustainable energy. IEEE Trans. Sustain. Energy **5**(2), 577–586 (2014)
15. Chademo: Dc fast charging
16. W. Tang, S. Bi, Y.J. Zhang, Online speeding optimal charging algorithm for electric vehicles without future information. In *IEEE International Conference on Smart Grid Communications*, pp. 175–180, IEEE (2013)
17. Z. Ma, D. Callaway, I. Hiskens, Decentralized charging control for large populations of plug-in electric vehicles. In *IEEE Annual Conference on Decision and Control*, pp. 206–212, IEEE (2010)
18. L. Gan, U. Topcu, S. Low, Optimal decentralized protocol for electric vehicle charging. IEEE Trans. Power Syst. **28**(2), 940–951 (2013)

Chapter 7
Hybrid Charging Control of Electric Vehicles

7.1 Introduction

Today's transportation section accounts for a significant portion of petroleum consumption and greenhouse gases emissions worldwide. Statistics show that 63.7% of the petroleum consumed in the world in 2012 was due to the transport sector, which caused emission of 7135 million tons of carbon dioxide into the environment [1]. The world's fossil fuels scarcity, as well as the growing environmental crisis associated with their wide usage, are driving the electrification of transportation and extensive use of electric vehicle (EVs). EVs are emerging as an efficient and sustainable alternative for private and public road transportation. Though widespread implementation of EVs may introduce a solution to the world fossil fuel shortage and air pollution concerns, the growing EV load also brings up multiple technical issues, such as voltage deviations, transformers and line saturations, increase of electrical losses. These issues may jeopardize the security and reliability of the power grid. As a consequence, intelligent charging and scheduling for EVs becomes a practical and important research problem.

A number of technical and regulatory issues, however, have to be resolved before the intelligent charging becomes a commonplace. The arrival of EVs and their required energy amount may appear to be random, which increase the demand-side uncertainties. The role of EV owners is also important in the interactions between the charging system and the EVs. From the EV owners' point of view, the degree of satisfaction should be an optimization objective. When departure, the EV owner hopes that energy in the battery remains as much as possible. In addition, EV owners may have various charging habits. Some are prone to individually determine their own charging profiles, while others may hope the charging system undertake the charging tasks for them. For instance, a future courier company may assign all the charging tasks of its driverless car fleet to a system controller of the charging park; meanwhile, some private car owners may prefer to control their charging patterns all by their own. And in many circumstances, these two kinds of user demands coexist. Therefore, a flexible and efficient EV charging mechanism has to be properly

© Springer Nature Singapore Pte Ltd. 2018

R. Wang et al., *Intelligent Microgrid Management and EV Control Under Uncertainties in Smart Grid*, https://doi.org/10.1007/978-981-10-4250-8_7

designed to dynamically coordinate the charging of EVs and satisfy the requirements of EV owners.

In this chapter, we consider the charging scheduling of a large number of EVs at a charging station. Stimulated by the fact that in practical scenarios, both centralized and decentralized charging architectures have their shortcomings and EV owners may have various charging preferences, a hybrid centralized–decentralized (HCD) EV charging mechanism is developed which offers flexible charging choices for customers. In this charging scheme, EV owners can either assign the charging tasks to system controller or individually choose the charging profiles based on their own preferences. In addition, the stochastic characteristics of EVs such as the arrival/departure times and charging demands are all taken into account. The main contributions of this chapter can be briefly summarized as follows:

- On the centralized charging control side, we formulate the coordinated EV charging problem into a convex optimization problem which aims at minimizing the charging cost of the whole EV fleet over a time horizon. To tackle with the system condition dynamics where EVs' arrival/departure times and their charging demands are uncertain, a model predictive control (MPC)-based scheduling mechanism is developed. The scheme allows more information of EV arrivals (departures) and charging demands to be effectively incorporated into the charging mechanism when such information is available.
- On the decentralized charging control side, a leader–follower noncooperative Stackelberg game is formulated to model the interactions between the system controller and EVs. The game aims at maximizing the profit of the charging system and utilities of the EV owners. We prove the existence of the generalized Stackelberg equilibrium (GSE) where both the leader and followers reach their equilibrium states. It is shown that the GSE also represents the socially optimal solution. Moreover, the communication burden between EVs and the system controller is low, and the proposed decentralized charging scheme is robust to poor communication channels.
- We further investigate the interactions between these two charging groups. It is shown that an optimal energy cap exists for the decentralized charging group which maximize the entire system's revenue. Moreover, an optimal energy allocation algorithm is proposed to find such energy cap.

The remainder of this chapter is organized as follows: Sect. 7.2 introduces the system model of the hybrid centralized–decentralized EV charging mechanism. In Sect. 7.3, we present the centralized charging control strategy where model predictive control (MPC)-based method is adopted to tackle with the system dynamics. In Sect. 7.4, a leader–follower noncooperative Stackelberg game-based decentralized EV charging scheme is introduced. The existence of equilibrium state and its optimality are analyzed. The simulation results and discussions are presented in Sect. 7.5. Finally, we conclude this chapter in Sect. 7.6.

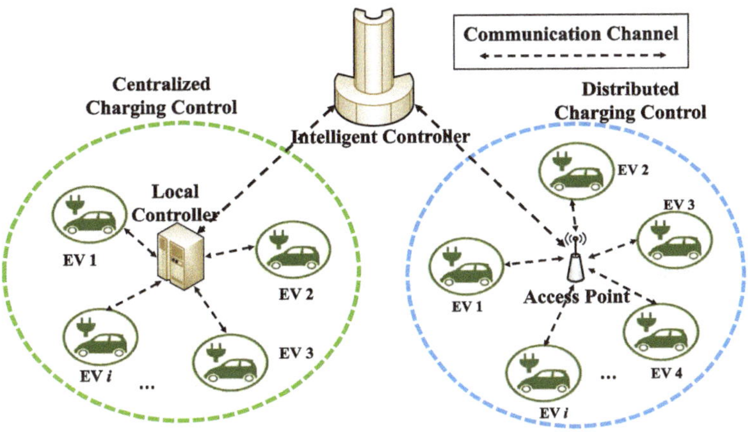

Fig. 7.1 Illustration of the system architecture

7.2 System Model

We consider an intelligent charging system (e.g., a charging park) which offers two charging options for the customers: (1) centralized charging: utilizing centralized infrastructure to collect information from EVs and centrally optimize EV charging considering the grid technical constraints; (2) decentralized charging: The vehicle owners directly control their EVs' charging patterns according to their own preferences. The general structure of the system is shown in Fig. 7.1. The system controller, the local controller(s), and EVs are the main players in this HCD charging control scheme. The particulars of the system operation and main principles associated with the modeling outlines of these components are explained in the following subsections.

7.2.1 Centralized Charging Control Model

A local controller is responsible for scheduling the charging patterns of a group of EVs on behalf of the their owners. If the number of EVs is large, the EVs can be classified into several groups (e.g., according to their geographical positions) and one local controller is responsible for the charging tasks of one EV group. The local controller and EVs (the local controller and system controller as well) are connected through two-way communication infrastructures (e.g., a local area network (LAN)). The operation time of the charging system is divided into discrete time intervals with equal length. The length of an interval is denoted by τ, which can vary from 5 min to half an hour based on the charging traffic conditions. Let $\mathcal{A}$ denote the set of EVs who participate in the centralized charging scheme. Adopting a general scheduling model, we define EV charging scheduling vector $\mathbf{x_a}$ and state vector $\mathbf{y_a}$ as follows:

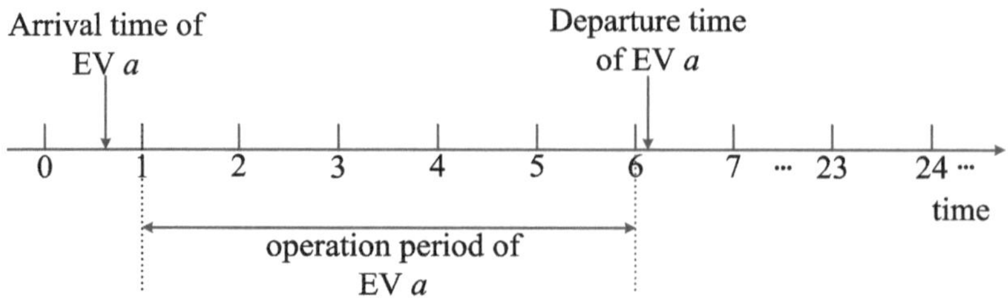

Fig. 7.2 Available charging period of electric vehicle a

$$\mathbf{x_a} = [x_a^1, x_a^2, \ldots, x_a^H] \text{ and } \mathbf{y_a} = [y_a^1, y_a^2, \ldots, y_a^H], \tag{7.1}$$

where $H \geq 1$ is the scheduling horizon which indicates the number of time slots ahead that are taken into account for decision-making in the EVs' charging scheduling. For each coming time slot $h \in \mathcal{H} = [1, 2, \ldots, H]$, let a binary variable $y_a^h = 0/1$ denote the state of EV a (not charging/charging) and a variable x_a^h denote the charging demand of EV a at time h. For each EV a with the maximum allowable charging rate R_a^{max} and the minimum charging rate R_a^{min}, we have

$$y_a^h \cdot R_a^{min} \cdot \tau \leq x_a^h \leq y_a^h \cdot R_a^{max} \cdot \tau. \tag{7.2}$$

Let t_a^s and t_a^f denote the plug-in time and departure time of EV a, respectively. Since we divide time into multiple discrete time slots, the available charging time of EV a, denoted by $\mathcal{T}_a$, is defined as the set of continuous time slots fall between the plug-in time t_a^s and plug-out time t_a^f, as depicted in Fig. 7.2. Obviously, we have $y_a^h = 0$ if $h \notin \mathcal{T}_a$. Further denote the battery capacity, initial battery energy, and desired departure battery energy of EV a by E_a^{cap}, E_a^s, and E_a^d, respectively. Obviously, we have $E_a^d \leq E_a^{cap}$. The desired departure state of charge (SOC) of EV a is defined as $\gamma_a^d = E_a^d / E_a^{cap}$, where $0 < \gamma_a^d \leq 1$. The local controller can automatically detect the arrival time t_a^s, battery capacity E_a^{cap}, and initial battery energy E_a^s of EV a when it connects to the charging plug. The departure time t_a^f and desired departure SOC γ_a^d are provided to the local controller by owner of EV a before the charging is began. Given t_a^s and t_a^f, the available charging period $\mathcal{T}_a$ can be easily obtained. Given the above descriptions, we have the following constraints intuitively:

$$E_a^s + \sum_{h \in \mathcal{T}_a} x_a^h \geq E_a^d \tag{7.3}$$

$$E_a^d \leq E_a^{cap} \tag{7.4}$$

Considering the fact that customers are risk averse, they would be reluctant to join the scheme if they face the financial risks associated with electricity price uncertainty (i.e., their EV may be charged during periods when the electricity prices are high).

Thus, it is assumed that the local controller offers flat electricity price p_c (which is announced in advance) for the EVs in return for their participation in this centralized charging scheme.

7.2.2 Decentralized Charging Control Model

Let $\mathcal{B}(h)$ denote the set of EVs engaging the decentralized charging scheme at time h. EVs are able to communicate with the system controller via two-way communication channels, as illustrated in Fig. 7.1. For a particular time slot $h \in \mathcal{H}$, the system controller has a limited energy E_m^h that it can provide to the $B(h)$ connected vehicles for charging, where $B(h) = |\mathcal{B}(h)|$. The system controller charges the EVs a price of p_d^h for one unit of electricity. For each EV $b \in \mathcal{B}(h)$, let x_b^h denote the amount of energy it requests from the charging system so as to meet its energy requirements. The energy demand x_b^h may vary for different EVs based on different parameters such as battery capacity E_b^{cap}, current SOC γ_b^c, desired plug-out SOC γ_b^d, the time varying electricity price p_d^h, and the travel plans (two identical EVs may have different travel plans and may have different energy demands). We assume that EVs will only request the amount of energy they currently needed subject to their immediate need for charging and EVs compete with each other for the limited scarce available energy. Thus, the following constraints must be satisfied for the total amount of energy EVs charged at time slot h:

$$\sum_{b \in \mathcal{B}(h)} x_b^h \leq E_m^h, \tag{7.5}$$

where E_m^h is the energy cap for the decentralized charging group. Obviously, the demand of the connected EVs is coupled through the above constraint. For the system controller, it tries to properly optimize the electricity price p_d^h such that the revenue for selling the energy is maximized. A lower electricity price means sacrificing revenues. However, if the price is set too high, customers (EVs) may reduce their demand, find alternative charging markets, or even wait until the price drops, also amounting to losing profits. Thus, a suitable p_d^h has to be decided to maximize the benefits of the charging system.

The interactions between system controller and EVs can be modeled as a leader–follower noncooperative Stackelberg game, in which there is a single leader (system controller) and multiple followers (EVs). The system controller chooses the total amount of energy it provides to EVs in $\mathcal{B}(h)$ and the electricity price. Given these two parameters, EVs respond to the controller by properly choosing their own charging demands. The game can be defined in its strategic form as

$$\mathcal{S} = \{(\mathcal{B}(h) \cup \{\text{system controller}\}), \{x_b^h\}_{b\in\mathcal{B}(h)}, E_m^h, p_d^h, \{U_b^h\}_{b\in\mathcal{B}(h)}, U_{sc}^h\}, \tag{7.6}$$

where U_b^h and U_{sc}^h are utility functions of EV b and system controller, respectively.

Note that the main task of the system controller is to properly coordinate the charging profiles of all the connected EVs (belong to either the centralized charging group or the decentralized charging group) to minimize the cost of the whole system.

7.3 Centralized Charging Scheme

7.3.1 Global Optimal Scheduling

To find global optimal EV charging profiles during the day, we first make the following assumptions: (1) The arrival time and departure time of each EV in the set $\mathcal{A}$ are known; (2) the plug-in SOC and desired plug-out SOC for each EV in the set $\mathcal{A}$ are known; (3) the local controller collects all the information accordingly and then performs the scheduling optimization. The local controller solves the following optimization problem to obtain the global optimal charging scheduling sequences:

$$\begin{aligned} \min_{\mathbf{X},\mathbf{Y}} \quad & \sum_{h=1}^{H} C_g^h(l^h) \\ \text{s.t.} \quad & l^h = \sum_{a \in \mathcal{A}} x_a^h \\ & x_a^h \geq 0, \ \ (7.2)\ (7.3) \text{ and } (7.4) \\ & y_a^h \in \{0, 1\} \text{ and } y_a^h = 0, \ \text{ if } h \notin \mathcal{T}_a, \end{aligned} \tag{7.7}$$

where $\mathbf{X} = [\mathbf{x_1}, \mathbf{x_2}, \ldots, \mathbf{x_a}, \ldots]^{\mathrm{T}}$ and $\mathbf{Y} = [\mathbf{y_1}, \mathbf{y_2}, \ldots, \mathbf{y_a}, \ldots]^{\mathrm{T}}$ are matrices of decision vectors $\mathbf{x_a}$ and $\mathbf{y_a}$ for $a \in \mathcal{A}$, respectively; $C_g^h(\cdot)$ is the cost function of the centralized charging system for generating or importing electricity, which is assumed to be an increasing convex function. The convex property reflects the fact that each additional unit of power needed to serve the demands is provided at a higher cost. Without loss of generality, we consider quadratic cost function $C_g^h(l^h) = b^h l^h + a^h {l^h}^2$ throughout this chapter [2, 3]. The global scheduling optimization problem can be interpreted as to minimize the total cost of the EV charging system during the day, by optimizing over the EV charging scheduling matrix $\mathbf{X}$ and state matrix $\mathbf{Y}$. Problem (7.7) is a mixed-integer quadratic programming (MIQP) problem, which can be effectively tackled by cutting plane method, branch and bounded method, etc. The solution to this problem provides the global (off-line) optimal EV charging scheduling sequences during the whole day. However, this scheduling scheme is impractical since the EVs' arriving and departing patterns are unknown and so are their parameters (current SOCs, plug-out SOCs). In the following subsection, we introduce a practical dynamic scheduling approach, which relaxes the assumptions adopted in the global optimal scheduling problem (7.7). The solution of this dynamic scheduling approach performs close to the global optimal scheduling scheme.

7.3.2 A Dynamic Scheduling Approach

One difficulty of the centralized charging lies in the fact that the system is dynamic with EVs coming and departing all the time. Thus, it is not possible to have a stationary long-term scheduling profile. To tackle with the system condition dynamics, we adopt the model predictive control (MPC) approach (also known as "receding horizon approach") [4, 5], of which the basic idea is to calculate the optimal control sequences yet to implement only the first step of them. In other words, the centralized EV scheduling problem is solved at time $h = \tau$ ($\tau \in \mathcal{H}$ denotes the current time index.) for the remaining horizon $[\tau, \tau + 1, \ldots, W^\tau]$, yet only the solution for the current time slot τ is implemented (W^τ is the decision-making horizon). In the next time slot, the local controller shall update the system information (e.g., the set of connected EVs currently, their current SOC, and desired plug-out SOC) and redo the calculations. The time horizon for the decision-making can be defined as the latest plug-out time of the EV in the current connected vehicle set, i.e., $W^\tau = \max_{a \in \mathcal{A}(\tau)} \lfloor t_a^f \rfloor$, where $\mathcal{A}(\tau)$ is the current connected vehicle set. The illustration of the current time horizon for the decision-making is depicted in Fig. 7.3. Mathematically, the optimization problem at current time τ can be formulated as:

$$
\begin{aligned}
\min_{x_a^h, y_a^h} \; & \sum_{h=\tau}^{W^\tau} C_g^h(l^h) \\
\text{s.t.} \; & l^h = \sum_{a \in \mathcal{A}} x_a^h \\
& x_a^h \geq 0, \; (7.2)\ (7.3) \text{ and } (7.4) \\
& y_a^h \in \{0, 1\} \text{ and } y_a^h = 0, \; \text{if } h \notin \mathcal{T}_a \\
& a \in \mathcal{A}(\tau), \; h \in [\tau, \tau + 1, \ldots, W^\tau].
\end{aligned}
\tag{7.8}
$$

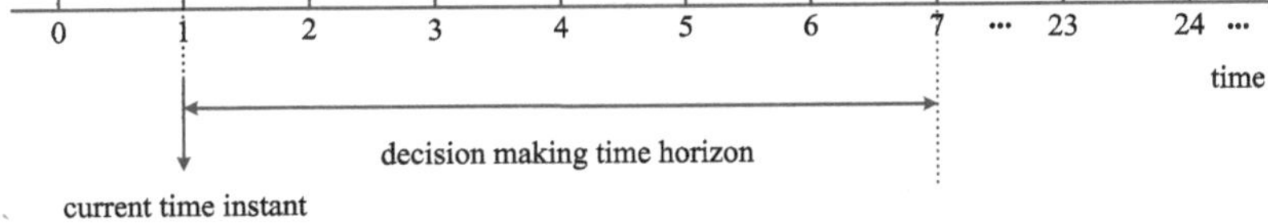

Fig. 7.3 The illustration of the current time horizon for the decision-making of the charging scheduling, i.e., $W^\tau = \max_{a \in \mathcal{A}(\tau)} \lfloor t_a^f \rfloor$. The current connected EVs' set is $\mathcal{A}(\tau) = \{\text{EV 1, EV 2, EV 3, EV 4, EV 5}\}$

The dynamic EV charging scheduling problem (7.8) at the beginning of time slot τ is still a MIQP problem which can be solved efficiently by many commercial optimization software including CPLEX, Mosek, FortMP, and Gurobi. By solving (7.8), we obtain charging scheduling sequences x_a^h, $a \in \mathcal{A}(\tau)$, $h \in [\tau, \tau+1, \ldots, W^\tau]$, among which only the charging scheduling sequences x_a^h, $a \in \mathcal{A}(\tau)$, $h = \tau$ are executed, and other scheduling sequences x_a^m, $a \in \mathcal{A}(\tau)$, $m \in [\tau+1, \ldots, W^\tau]$ are discarded, which will be finally updated at the beginning of time slot m.

7.4 Decentralized Charging Scheme

7.4.1 Game Formulation

In this section, we formulate the interactions between system controller and EVs into a leader–follower noncooperative Stackelberg game, where the system controller acts as the leader and the EVs are followers. At any time slot h, two principle components of the game $\mathcal{S} = \{(\mathcal{B}(h) \cup \{\text{system controller}\}), \{x_b^h\}_{b\in\mathcal{B}(h)}, E_m^h, p_d^h, \{U_b^h\}_{b\in\mathcal{B}(h)}, U_{sc}^h\}$ are the utility functions of the leader (system controller) U_{sc}^h and the followers (EVs) U_b^h, $b \in \mathcal{B}(h)$. We have detailed discussions as follows:

7.4.1.1 Utility Functions of EVs

EV's utility function captures the benefit it obtains for consuming the demand energy. The utility function $U_b^h(x_b^h, \mathbf{x}_{-b}^h, \alpha_b^h, \beta_b^h, p_d^h)$ of EV b is defined as a function of the energy it charges. Here, x_b^h is the requested charging energy of b from the charging station. $\mathbf{x}_{-b}^h$ is the vector formed of all players decision variables except the one of player b, i.e., $\mathbf{x}_{-b}^h = (x_1^h, x_2^h, \ldots, x_{b-1}^h, x_{b+1}^h, \ldots)$. $\alpha_b^h > 0$ and $\beta_b^h > 0$ are parameters measuring the charging habit of EV b. The value of α_b^h and β_b^h may depend on the current SOC, the battery capacity, and the travel plan of the EV b. In addition, the price of electricity p_d^h also influences the charging benefit of a EV. Mathematically, we have the following assumptions on the properties of the utility function of EV b:

(i) Assumption 7.1: The utility functions U_b^h is non-decreasing with respect to the amount of energy the EV charges. In other words, each EV tends to charge more if possible unless it reaches its maximum battery level, i.e.,

$$\frac{\partial U_b^h(x_b^h, \mathbf{x}_{-b}^h, \alpha_b^h, \beta_b^h, p_d^h)}{\partial x_b^h} \geq 0. \tag{7.9}$$

(ii) Assumption 7.2: An EV has a declining marginal benefit with respect to the charging amount. This statement can be interpreted from these two aspects: (1) The marginal charging time (i.e., drivers' waiting time) increases since the charging rate slows down when the battery gets drenched; (2) satisfaction level of an EV gradually gets saturated when more and more energy is consumed; i.e.,

$$\frac{\partial^2 U_b^h(x_b^h, \mathbf{x}_{-b}^h, \alpha_b^h, \beta_b^h, p_d^h)}{\partial {x_b^h}^2} \leq 0. \tag{7.10}$$

(iii) Assumption 7.3: EV's benefit gets lower when the electricity price increases, i.e.,

$$\frac{\partial U_b^h(x_b^h, \mathbf{x}_{-b}^h, \alpha_b^h, \beta_b^h, p_d^h)}{\partial p_d^h} < 0. \tag{7.11}$$

Without loss of generality, the quadratic utility function is defined as:

$$\begin{aligned} &U_b^h(x_b^h, \mathbf{x}_{-b}^h, \alpha_b^h, \beta_b^h, p_d^h) \\ &= -\frac{1}{2}\alpha_b^h (x_b^h)^2 + \beta_b^h \cdot x_b^h - p_d^h \cdot x_b^h \end{aligned} \tag{7.12}$$

Note that the game formulation we proposed in this chapter is a general methodology which is not restricted to the current quadratic utility function. As long as the utility function satisfies the above assumptions, the proposed method can be applied with virtually no change.

7.4.1.2 Utility Function of the System Controller

The objective of the system controller is to maximize the revenue for selling the electricity to EVs, and thus, the utility function of the system controller is defined mathematically as:

$$\begin{aligned} U_{sc}^h = p_c \cdot \sum_{a \in \mathcal{A}(h)} x_a^h + p_d^h \cdot \sum_{b \in \mathcal{B}(h)} x_b^h \\ -C_m^h \left(\sum_{a \in \mathcal{A}(h)} x_a^h + \sum_{b \in \mathcal{B}(h)} x_b^h \right), \end{aligned} \tag{7.13}$$

where C_m^h is the cost function of the charging system. Utility function U_{sc}^h captures the revenue for selling the energy (fist two terms) and the cost for generating or buying the energy (the last term). Without loss of generality, we also consider the quadratic cost function here, i.e., $C_m^h(x) = n^h x + m^h x^2$. In the proposed game, the system controller can control the price for selling the energy p_d^h and total energy cap E_m^h. The EVs respond to this price and choose the amount of energy to charge x_b^h to maximize their utilities, and simultaneously, they have to ensure that their total charging demand should not exceed the energy cap E_m^h. Note that the centralized charging scheduling sequences x_a^h, $a \in \mathcal{A}(h)$ are determined by the centralized charging scheme (local controller). In this regard, for a fixed electricity price p_d^h, an EV b solves the following optimization problem:

$$\max_{x_b^h} \; U_b^h(x_b^h, \mathbf{x}_{-b}^h, \alpha_b^h, \beta_b^h, p_d^h) \tag{7.14}$$

$$\text{s.t.} \sum_{b \in \mathcal{B}(h)} x_b^h \leq E_m^h. \tag{7.15}$$

Obviously, the charging strategy of EV b not only depends on its own utility function but also depends on other EVs' charging strategies through constraint (7.15), and this constraint is shared by all the players (i.e., EVs). This game is a jointly convex generalized Nash equilibrium problem[1,2,3] (GNEP) due to the same shared "coupled constraint" (7.15) and the max-concave (i.e., min-convex) objective functions of EVs [6]. Then, after all the EVs' charging amount reaches the GNE, the system controller optimize the energy price p_d^h to maximize the revenue of the system. Given the GNE charging amount of EVs $(x_b^h, \mathbf{x}_{-b}^h)$, the system controller solves the following problem:

$$\max_{p_d^h} U_{sc}^h \tag{7.16}$$

to maximize the system revenue. The solution of the formulated noncooperative leader–follower generalized Stackelberg game (GSG) is the generalized Stackelberg equilibrium (GSE) in which the leader finds its optimal price and the followers reach their equilibrium states. At this equilibrium, no player (i.e., both the leader and the followers) can increase his utility by changing unilaterally his strategy to any other feasible point. Here, we term the formulated game as generalized Stackelberg game (GSG) rather than Stackelberg game because of the coupled constraint (7.15) for the followers. Since the followers' strategies are coupled, they need to seek a GNE instead of a traditional Nash equilibrium (NE). Therefore, the formulated game is termed as GSG whose solution is called GSE.

---

[1]The generalized Nash equilibrium problem (GNEP) is a noncooperative game in which each player's admissible strategy set depends on the other players' strategies.

[2]In a noncooperative game, if the players' actions are coupled solely through the constraints, then this game is a special class of game whose solution is a generalized Nash equilibrium (GNE).

[3]The objective functions of EVs are all min-convex (max-concave) functions, and the strategy set which is constrained by a single linear function is closed and convex with respect to all variables; then, we have that this formulated GNEP is jointly convex [6]. Detailed discussions will be presented in the next section.

7.4.2 Existence of GSE

We first specify the definition of GSE and then discuss in detail the existence and the properties of it.

Definition 7.1 For the GSG formulation

$$\mathcal{S} = \{(\mathcal{B}(h) \cup \{\text{system controller}\}), \{x_b^h\}_{b\in\mathcal{B}(h)}, E_m^h, p_d^h, \{U_b^h\}_{b\in\mathcal{B}(h)}, U_{sc}^h\}$$

defined in Sect. 7.4.1, where U_{sc}^h and $U_b^h, b \in \mathcal{B}(h)$ are utility functions of the leader and followers given by (7.13) and (7.12), respectively. A strategy set $(\mathbf{x}^{h^*}, p_d^{h^*})$ constitutes the GSE of the game, if and only if the following inequalities are satisfied:

$$U_b^h(x_b^{h^*}, \mathbf{x}_{-b}^{h^*}, \alpha_b^h, \beta_b^h, p_d^{h^*}) \geq U_b^h(x_b^h, \mathbf{x}_{-b}^{h^*}, \alpha_b^h, \beta_b^h, p_d^{h^*})$$
$$\forall x_b^{h^*} \in \mathbf{x}^{h^*}, b \in \mathcal{B}(h), \sum_{b\in\mathcal{B}(h)} x_b^h \leq E_m^h$$

and

$$U_{sc}^h(p_d^{h^*}, \mathbf{x}^{h^*}) \geq U_{sc}^h(p_d^h, \mathbf{x}^{h^*}). \tag{7.17}$$

In other words, no EV can increase its revenue by deviating from its GSE charging amount $\mathbf{x}^{h^*}$ and no price other than the GSE price $p_d^{h^*}$ can improve the utility of the charging system.

Typically, in noncooperative games, the existence of Nash equilibrium is not always guaranteed. For the followers' game, to investigate the existence of GNE in response to a price p_d^h, we first propose the following definitions and theorems:

Definition 7.2 We say a game satisfies the convexity assumption if the following condition holds: For every player $v \in \mathcal{N}$ and every strategy $x_v \in \mathbb{R}^{n_v}$, where $\mathcal{N}$ is the set of players, the objective function $U_v(\cdot, x_v, \mathbf{x}_{-v})$ is min-convex (max-concave) and the strategy set $X_v(\mathbf{x}_{-v})$ is closed and convex. Note that we use $X_v(\mathbf{x}_{-v})$ to represent the strategy set of player v since his strategy set is dependent on other players' strategies.

Obviously, in the proposed followers' game, for each players, the objective function U_b^h is max-concave and the strategy set which is merely confined by constraint (7.15) is closed and convex. Thus, the followers' game satisfies convexity assumption.

Definition 7.3 Let a GNEP be given, which satisfies convexity assumption, this GNEP is jointly convex if for some closed convex $\mathbf{X} \subseteq \mathbb{R}^n$ $(n = n_1 + n_2 + \cdots + n_N)$ and all $v \in \mathcal{N}$, we have

$$X_v(\mathbf{x}_{-v}) = \{x_v \in \mathbb{R}^{n_v} : (x_v, \mathbf{x}_{-v}) \in \mathbf{X}\}. \tag{7.18}$$

For the proposed followers' game, it is easy to check that the strategy set of EV b is:

$$X_b(\mathbf{x}_{-b}^h) = \left\{ x_b^h \in \mathbb{R}_0^+, \sum_{b \in \mathcal{B}(h)} x_b^h \leq E_m^h \right\}. \tag{7.19}$$

Obviously, this game satisfies the jointly convex condition. Based on the previous definitions, the following theorem is proposed.

Theorem 7.1 *In a jointly convex GNEP, the utility function of each player U_v is continuously differentiable, and then, every solution of the variational inequality problem VI($\mathbf{X}$, $\mathbf{F}$)[4] is also a solution of GNEP, where $\mathbf{X}$ is as defined in the definition of jointly convex and $\mathbf{F} = [\frac{\partial U_v}{\partial x_v}]_{v=1}^N$.*

The proof for this theorem can be found in [6]. Note that Theorem 7.1 does not say that any solution of a jointly convex GNEP is also a solution of the VI($\mathbf{X}$, $\mathbf{F}$) and some solutions may lost. We further have the definition of the variational equilibrium (VE) as follows.

Definition 7.4 In a jointly convex GNEP, the utility function of each player U_v is continuously differentiable, and we call a solution of the GNEP that is also a solution of VI($\mathbf{X}$, $\mathbf{F}$) a variational equilibrium (VE).

In a GNEP, the existence of VE is of particularly interest since a VE is more socially stable than other GNE (if there exists any), and thus, it is a desirable equilibrium state [7]. Next, we will prove the existence and uniqueness of VE in our proposed followers' game.

Theorem 7.2 *If $\mathbf{X}$ is a compact convex set and $\mathbf{F}(\mathbf{x})$ is continuous on $\mathbf{X}$, then the variational inequality problem admits at least one solution $\mathbf{x}^*$.*

The proof for this theorem is lengthy and can be found in [8]. Considering the proposed followers' game, the strategy set of EVs

$$\mathbf{X}^h = \left\{ (x_1^h, x_2^h, \ldots, x_b^h, \ldots) : \right. \tag{7.20}$$

$$\left. \forall b \in \mathcal{B}(h),\ x_b^h \geq 0,\ \sum_{b \in \mathcal{B}(h)} x_b^h \leq E_m^h \right\}$$

is a Polyhedron, which is compact and convex. For the corresponding

$$\mathbf{F}^h = -\left[\frac{\partial U_b^h}{\partial x_b^h} \right]_{b=1}^{B(h)} = \begin{bmatrix} \alpha_1^h x_1^h + p_d^h - \beta_1^h \\ \alpha_2^h x_2^h + p_d^h - \beta_2^h \\ \vdots \\ \alpha_{B(h)}^h x_{B(h)}^h + p_d^h - \beta_{B(h)}^h \end{bmatrix} \tag{7.21}$$

---

[4]The variational inequality problem VI($\mathbf{X}$, $\mathbf{F}(\mathbf{x})$) consists in finding a vector $\bar{\mathbf{x}} \in \mathbf{X}$ such that $(\mathbf{y} - \bar{\mathbf{x}})^T \cdot \mathbf{F}(\bar{\mathbf{x}}) \geq 0$ for all $\mathbf{y} \in \mathbf{X}$.

is obviously continuous (linear), therefore we claim there exists VE in the followers' game. To investigate the uniqueness of VE, we propose the following theorem.

Theorem 7.3 *In a variation inequality problem VI($\mathbf{X}$, $\mathbf{F}$), if $\mathbf{F}(\mathbf{x})$ is strictly monotone on $\mathbf{X}$. Then the solution is unique, if one exists.*

The proof for this theorem is presented in the **Appendix** *C. Now turn to the definition of $\mathbf{F}^h$, we have that the Jacobian of $\mathbf{F}^h$ is*

$$\mathbf{JF}^h = \begin{bmatrix} \alpha_1^h & 0 & \dots & 0 \\ \cdot & \alpha_2^h & \dots & \cdot \\ \cdot & \cdot & \dots & \cdot \\ \cdot & \cdot & \dots & \cdot \\ 0 & 0 & \dots & \alpha_{B(h)}^h \end{bmatrix}, \tag{7.22}$$

which is a diagonal matrix with all the diagonal elements positive. In other words, $\mathbf{JF}^h$ is a positive-definite matrix so $\mathbf{F}^h$ is strictly monotone on $\mathbf{X}^h$. Therefore, given an electricity price p_d^h, there exists GNE and more precisely, an unique VE for the followers' GNEP.

Theorem 7.4 *For a fixed electricity price p_d^h, the unique VE is the socially optimal solution of the proposed followers' GNEP between EVs.*

The proof for Theorem 7.4 is presented in the **Appendix** *C. This theorem states that by solving the VI($\mathbf{X}^h$, $\mathbf{F}^h$), where $\mathbf{X}^h$ and $\mathbf{F}^h$ are defined by (7.20) and (7.21), respectively, the socially optimal solution of the followers' GNEP can be obtained. As a result, when the system controller sets its optimal price in response to the VE demand of the EVs, the GSG reaches its GSE, which represents the socially optimal solution.*

7.4.3 Solution and Algorithm

7.4.3.1 VE for the Followers' GNEP

For the decentralized charging scheme proposed in this chapter, the GNEP among the EVs is transformed into a strictly monotone variational inequality (VI) problem whose solution leads to the socially optimal VE. Numerous methods have been proposed to solve the VI problem, including projection method, relaxation method, decomposition method. In this chapter, we adopt Solodov and Svaiter (S–S) method to solve the VI problem [9, 10]. The S–S method is a kind of extragradient method (a subclass of the projection method) which can solve the VI problem efficiently. The S–S method works briefly as follows: Suppose $\mathbf{x}^k \in \mathbf{X}$ be the current approximation of the solution of VI($\mathbf{X}$, $\mathbf{F}$); first, we compute the point $P_X(\mathbf{x}^k - \mu^k \mathbf{F}(\mathbf{x}^k))$, where $P_X(\cdot)$ denotes the orthogonal projection map onto $\mathbf{X}$ and μ^k is a judiciously chosen positive steplength. Here, $P_X(\mathbf{x}^k - \mu^k\mathbf{F}(\mathbf{x}^k))$ is the solution of the following quadratic programming problem

$$\min_{\mathbf{x}\in\mathbf{X}} \frac{1}{2}\mathbf{x}^T\mathbf{x} - (\mathbf{x}^k - \mu^k\mathbf{F}(\mathbf{x}^k))^T\mathbf{x}; \tag{7.23}$$

next, the line segment between $\mathbf{x}^k$ and $P_X(\mathbf{x}^k - \mu^k\mathbf{F}(\mathbf{x}^k))$ is searched for a point $\mathbf{z}^i$ such that the hyperplane

$$\partial H_k = \{\mathbf{x} \in \mathbb{R}^n | < \mathbf{F}(\mathbf{z}^k), \mathbf{x} - \mathbf{z}^k >= 0\} \tag{7.24}$$

strictly separates $\mathbf{x}^k$ from the solution of the VI($\mathbf{X}, \mathbf{F}$) $\mathbf{x}^*$, where $< \cdot, \cdot >$ is the usual inner product in $\mathbb{R}^n$. To compute $\mathbf{z}^k$, an Armijo-type procedure is adopted, i.e., $\mathbf{z}^k = \mathbf{x}^k - \eta^k r(\mathbf{x}^k, \mu^k)$ where $\eta^k = \gamma^{\bar{i}}\mu^k$ with $\bar{i}$ is the smallest nonnegative integer i satisfying

$$\begin{aligned} < \mathbf{F}(\mathbf{x}^k - \gamma^i\mu^k r(\mathbf{x}^k, \mu^k)), r(\mathbf{x}^k, \mu^k) > \\ \geq \frac{\sigma}{\mu^k}\|r(\mathbf{x}^k, \mu^k)\|^2 \end{aligned} \tag{7.25}$$

where $r(\mathbf{x}^k, \mu^k) = \mathbf{x}^k - P_X(\mathbf{x}^k - \mu^k\mathbf{F}(\mathbf{x}^k))$ being the projected residual function; after the hyperplane ∂H_k is constructed, the next iteration $\mathbf{x}^{k+1}$ is computed by projecting $\mathbf{x}^k$ onto the intersection between the feasible set $\mathbf{X}$ with the halfspace $\partial H_k = \{\mathbf{x} \in \mathbb{R}^n | < \mathbf{F}(\mathbf{z}^k), \mathbf{x} - \mathbf{z}^k > \leq 0\}$ which contains the solution set $\mathbf{X}^*$. The details of the S–S method are shown in Algorithm 7.1. Upon solving the VI($\mathbf{X}, \mathbf{F}$), the VE demand of each EV can be obtained. Next, we show how to optimize the electricity price by the system controller given the VE of the EVs.

7.4.3.2 Electricity Price Optimization

To investigate the electricity price optimization, we first consider the Karush–Kuhn–Tucker (KKT) optimal condition system of the VI problem, which is given by

$$\mathbf{F}^h + \nabla_{\mathbf{x}}^h \left(\sum_{b\in\mathcal{B}(h)} x_b^h - E_m^h \right) \cdot \lambda = 0, \tag{7.26}$$

$$\lambda \left(\sum_{b\in\mathcal{B}(h)} x_b^h - E_m^h \right) = 0, \tag{7.27}$$

for some multiplier $\lambda \geq 0$. Note that if

$$\sum_{b\in\mathcal{B}(h)} x_b^h < E_m^h, \tag{7.28}$$

then some EVs are able to increase their charging demand to gain higher utilities. Finally, this constraint becomes an equality, and hence, at the VE,

$$\sum_{b \in \mathcal{B}(h)} x_b^h = E_m^h, \tag{7.29}$$

i.e., for a fixed electricity price, the sum of demands of all the EVs at the VE are equal to the available energy cap E_m^h. Note that in the game formulation, energy is assumed to be a scarce resource. The energy cap E_m^h should be lower than the total energy consumption capacity of the connected EVs. This avoids the trivial case where all the EVs get the energy allocation equal to their maximum capacity. From (7.26), we have that:

$$\lambda + \alpha_b^h x_b^{h^*} + p_d^h - \beta_b^h = 0, \tag{7.30}$$

for any $b \in \mathcal{B}(h)$. Thus, the electricity price should satisfy

$$p_d^h = \beta_b^h - \lambda - \alpha_b^h x_b^{h^*}. \tag{7.31}$$

Considering the utility function of the system controller U_{sc}^h from (7.13), obviously when p_d^h reaches its maximum the system can obtain the maximum utility, and therefore, the optimal price of the proposed game is:

$$p_d^{h^*} = \beta_b^h - \alpha_b^h x_b^{h^*}, \tag{7.32}$$

i.e., $\lambda = 0$ when the GSG reaches the GNE. The requested charging amount of each vehicle should be

$$x_b^{h^*} = \frac{\beta_b^h - p_d^{h^*}}{\alpha_b^h} \tag{7.33}$$

with the optimal electricity price $p_d^{h^*}$. This is the equilibrium state of the game.

7.4.3.3 Algorithm Design

In order to reach the equilibrium, the system controller and the EV have to communicate with one another to make their choices. Upon any EV b is plugged in, the system controller receives its utility parameters α_b^h and β_b^h via communication channels (e.g., V2G). The algorithm starts with the setting of energy cap E_m^h. Given the fixed amount E_m^h, the system controller solves VI($\mathbf{X}^h$, $\mathbf{F}^h$) to obtain the optimal charging strategy vector $\mathbf{x}^{h*}$ using the S–S mechanism. The system controller then gets the optimal electricity price p_d^{h*} adopting (7.32). p_d^{h*} is broadcasted to EVs through communication channels, and EVs determine their charging demand through solving (7.14), which is actually given by (7.33). Note that this algorithm shows it advantages over that in reference [11] in the following aspects: (1) Algorithm 7.2 is implemented in a distributed fashion (each EV chooses its own charging demand), and EVs undertake very low computational burden since the VI problem is solved by the system controller, while in [11], EVs have to participate into the problem solving of the VI problem. (2) By adopting our proposed algorithm, the communication traffic between the EVs and the system controller is very low. In each time slot, only one round-trip communication is implemented (i.e., the EVs submit their utility functions, and the system controller broadcasts the optimal electricity price), while in [11], dozens of round-trip messages have to be exchanged before the game reaches its GSE. Particularly, when the communication channel is poor, our approach can easily overcome the unstable channel by retransmission. But in [11], the game is difficult to reach GSE in this scenario. Since to implement the algorithm, it has to ensure that in each iteration, all the exchanged messages between the controller and EVs are available and correct. This would become more challenging when the EV fleet is of a large size. The details of the S–S scheme to find GNE and the proposed algorithm to reach GSE are depicted in Algorithms 7.1 and 7.2, respectively.

Algorithm 7.1 Solodov and Svaiter (S–S) method [9, 10]

Input: The matrix $\mathbf{F}^h$ and the strategy set $\mathbf{X}^h$ which are given in (7.20) and (7.21), respectively; Initial electricity price p_d^{h0}; Final tolerance ε.
Output: Optimal charging strategy vector $\mathbf{x}_b^{h*}$.

```
Begin
choose x^0 ∈ X^h, η_{-1} > 0, γ ∈ (0, 1), σ ∈ (0, 1), θ > 1, k = 0, gap = e, where e is a vector
  with entries equal to 1;
if ||gap|| < ε
then stop;
else
  compute μ^k = min{θ · η_{k-1}, 1};
  if r(x^k, μ^k) = x^k − P_X(x^k − μ^k F(x^k))
  then x^k ∈ X^{h*}, stop;
  else
  compute ī = arg min_{i∈Z+} {< F^h(x^k − γ^i μ^k r(x^k, μ^k)), r(x^k, μ^k) >≥ (σ/μ^k) ||r(x^k, μ^k)||^2},
    where η_k = γ^ī μ^k;
   compute z^k = x^k − η_k r(x^k, μ^k);
   compute the halfspace ∂H_k = {x ∈ ℝ^n | < F(z^k), x − z^k >≤ 0};
   compute x^{k+1} = P_{X^h ∩ H_k}(x^k);
   gap = x^{k+1} − x^k;
   k = k + 1;
   go to 3;
  end if
end if
End
```

Algorithm 7.2 Algorithm to reach GSE

Input: Utility function U_b^h for each vehicle $b \in \mathcal{B}(h)$; Initial electricity price p_d^{h0}.
Output: Optimal electricity price p_d^{h*}; optimal charging strategy x_b^{h*} selected by each vehicle $b \in \mathcal{B}(h)$.

Begin
Each EV $b \in \mathcal{B}(h)$ submits its utility function parameters α_b^h and β_b^h;
The system controller determines the energy cap E_m^h for the uncontrolled EVs;
The system controller solves VI($\mathbf{X}^h$, $\mathbf{F}^h$) by adopting Algorithm 7.1 and obtains the optimal charging strategy vector $\mathbf{x}^{h*}$;
The system controller computes the optimal electricity price p_d^{h*} based on (7.32);
The system controller broadcast the electricity price p_d^{h*} to all the EVs that are uncontrolled.
Each vehicle chooses their charing demand by solving problem (7.14) and obtains the optimal charging strategy x_b^{h*}, $b \in \mathcal{B}(h)$;
End

7.4.4 Algorithm to Determine a Proper E_m^h

In the distributed charging scheme, EVs compete with each other for a fair allocation of the scarce energy. Intuitively, when the energy cap E_m^h is low, the competition between EVs becomes fierce and the optimal energy price p_d^h gets high. In contrast, if the E_m^h is high, then p_d^h is low. Under both cases, the total revenue of the system U_{sc}^h is poor. Hence, we may assume that U_{sc}^h will first increase and then decline with respect to E_m^h (i.e., quasi-concave) and a proper E_m^h exists which can maximize U_{sc}^h (such assumption will be proved in the following simulation part). Various algorithms can be adopted to search the optimal E_m^h, including genetic algorithm (EA), Newton–Raphson method, gradient descent method. In this chapter, we assume U_{sc}^h is derivable

Algorithm 7.3 Algorithm to search an optimal E_m^h

Input: Starting point E_m^{h0}, tolerance ε;
Output: Optimal energy cap E_m^{h*}.

Begin
$x_0 = E_m^{h0}, k = 0$;
compute $y_0 = U_{sc}^h(x_0)$ based on Algorithm 2;
while (true)
  compute $\nabla U_{sc}(x_k) = \frac{U_{sc}(x_k+\epsilon)-U_{sc}(x_k)}{\epsilon}$, where ϵ is a small number;
  $x_k = x_k + \alpha \cdot \nabla U_{sc}(x_k)$;
  $y_1 = U_{sc}^h(x_k)$;
  if $\|y_1 - y_0\| \leq \varepsilon$
    break;
  end if
  $y_0 = y_1$;
  $k = k + 1$;
end while
$E_m^{h*} = x_k$;
End

with respect to E_m^h and propose the following algorithm based on gradient descent method to search an optimal E_m^h. The effectiveness of the proposed algorithm will be verified in the following section.

7.5 Experimental Evaluation

In this section, we present simulation results based on real-world traces for assessing the performance of the proposed mixed EV charging scheme and evaluate the effects of different parameters.

7.5.1 *Simulation Setting*

In this chapter, the units of the electricity price, the cost functions, and the utility functions are US cent ¢/KWh. For the centralized charging, the scheduling horizon is 8 h with time evenly divided into 32 time slots, i.e., the length of each time slot is 15 mins. The number of connected vehicles currently is 100 unless otherwise stated. The plug-out times are uniformly distributed between 1 time slot and 32 time slots. The amounts of energy needed for the EVs are evenly distributed between 8 and 64 KWh. The maximum allowable charing rate of an EV is 28 KW, and the minimum charging rate of an EV is 0 KW. For the cost function of the electricity acquisition $C_g^h(\cdot)$, we set $a^h = 1 \times 10^{-3}$ ¢/KWh^2 and $b^h = 1.6$ ¢/KWh. To solve the optimization problem (7.8), interior point method is adopted, which can solve the convex optimization problem efficiently. For the distributed charging, 100 vehicles participate in this scheme. Unless otherwise stated, their utility function parameters α_b^h and β_b^h are chosen randomly in the range of [0.75, 1.25] and [13, 15], respectively. The energy cap E_m^h is set as 700 KWh by default. Note that all statistical results are averaged over all possible random values of the EVs' parameters using 500 independent simulation results.

7.5.2 *Results and Discussion*

We first investigate how the optimal charging price p_d^{h*} varies with respect to energy cap E_m^h. The energy cap E_m^h is linearly varied from 675 to 725 KWh for different EV numbers $B(h) = 95$, 100, and 105. Adopting Algorithm 7.2, we compute the corresponding optimal electricity price p_d^{h*}. The results are depicted in Fig. 7.4. It is obviously shown that the average optimal price decreases with the energy cap. This is due to the fact that when the total available capacity of the charging system increases, the grid has more energy to sell, and thus, the competition between EVs gets weaker and price declines. In other words, as the available energy increases, the

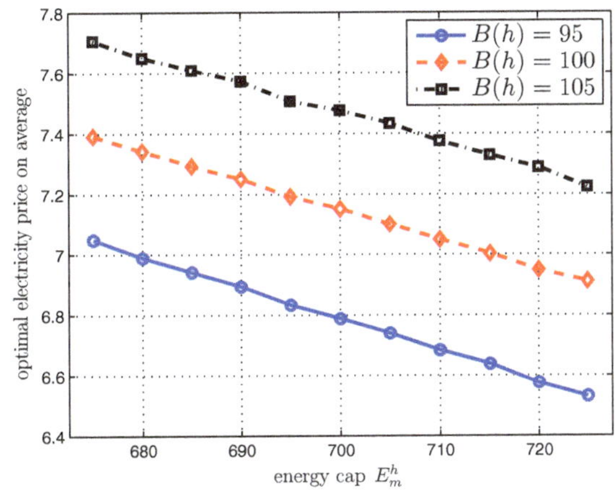

Fig. 7.4 The variation of optimal electricity price for decentralized controlled vehicles p_d^{h*} with respect to their energy cap E_m^h

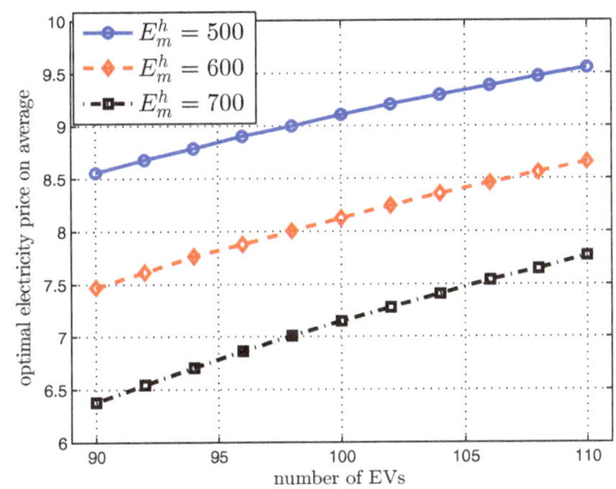

Fig. 7.5 The variation of optimal electricity price for decentralized controlled vehicles p_d^{h*} with respect to their number

system controller has to reduce the energy price to encourage the vehicle to charge more energy. Meanwhile, in Fig. 7.5, the effect of the number of connected EVs on the average optimal electricity price is presented. It appears that the growing vehicle number leads to a growth in the average optimal price. The reason is that a larger vehicle number means increasing in electricity demand. In this case, the system controller can set a higher electricity price to stimulate EVs charge less energy.

The impacts of EVs' utility function parameters α_b^h and β_b^h on the average optimal electricity price are illustrated in Figs. 7.6 and 7.7, respectively. To do the test, we vary the value of α_b^h for different ranges of $\beta_b^h \in [10, 12]$, $[12, 14]$, and $[14, 16]$, respectively. The test to assessing the impact of β_b^h is conducted in a similar way, i.e., β_b^h is increased for various ranges of $\alpha \in [0.5, 0.8]$, $[0.8, 1.1]$, and $[1.1, 1.4]$, respectively. We observe that the optimal price is a decreasing function of α_b^h. In contrast, a rise in β_b^h presents an upper trend in the optimal price. The reason is that a higher α_b^h indicates that the EV's marginal utility declines. Thus, EVs are prone

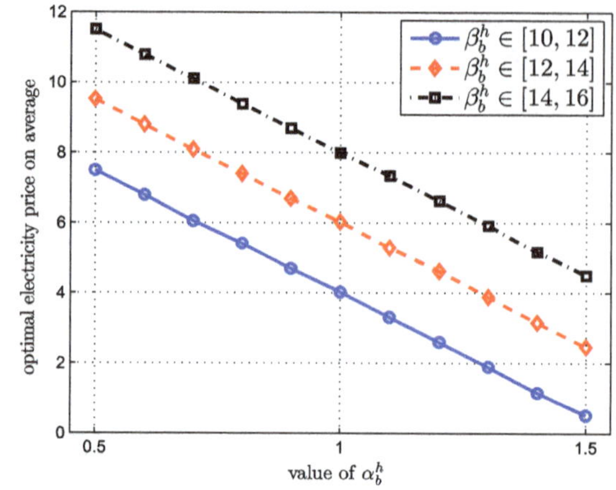

Fig. 7.6 The variation of optimal electricity price for decentralized controlled vehicles p_d^{h*} with respect to the users' utility parameter α_b^h

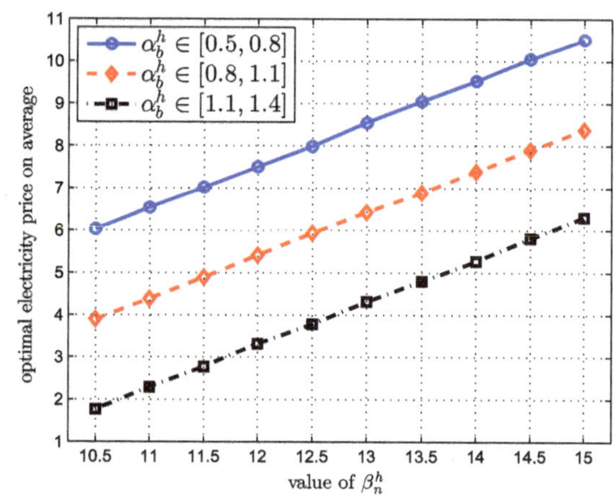

Fig. 7.7 The variation of optimal electricity price for decentralized controlled vehicles p_d^{h*} with respect to the users' utility parameter β_b^h

to charge less, and the corresponding electricity price decreases. While on opposite, an increment on β_n^h implies a rise of the marginal utility of the vehicle and therefore leads to a brisker energy demand and a higher electricity price. These results also verify the theoretical analysis result presented in Sect. 7.4.3.

In Fig. 7.8, we present the total average utilities of the charging system as a function of the energy cap of the distributed charging scheme E_m^h. To do the test, 100 vehicles are centrally controlled and the other 100 vehicles choose their charging profiles by their own. Energy cap of the latter group E_m^h increases from 300 to 700 KWh, and we compare the average utilities of the whole system. It appears that the average utility first shows an upper trend and then declines. There exists an optimal E_m^h which maximize system's utility. Thus, by adopting Algorithm 7.3 proposed in the previous section, the system controller can properly determine an optimal E_m^h to maximize its revenue given the system condition. The cost function C_m^h is further altered to investigate its impact. We observe that if the cost for acquiring knowledge grows,

Fig. 7.8 The utility of the system with respect to decentralized controlled EVs' energy cap E_m^h under different cost functions

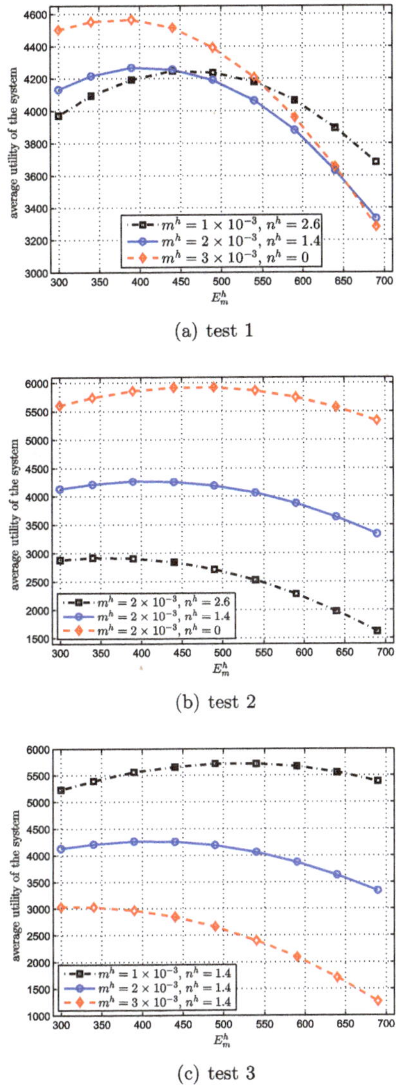

i.e., parameters m^h or n^h increase, both system utility and optimal energy cap E_m^{h*} decline. For the centralized controlled EV group, since the charging requirements have to be satisfied, the adjustment conducted by the system controller is relatively limited. Therefore, as the energy cost increases, the system controller is prone to cut down the proportion of energy allocated to decentralized controlled EVs so that it can curtail the energy expenses.

In Fig. 7.9, we evaluate how the vehicle numbers in both charging groups impact the system utility. To conduct this test, the total connected vehicles are fixed to 200,

Fig. 7.9 The utility of the system with respect to the number of vehicles that are centralized controlled (the total number of EVs is 200)

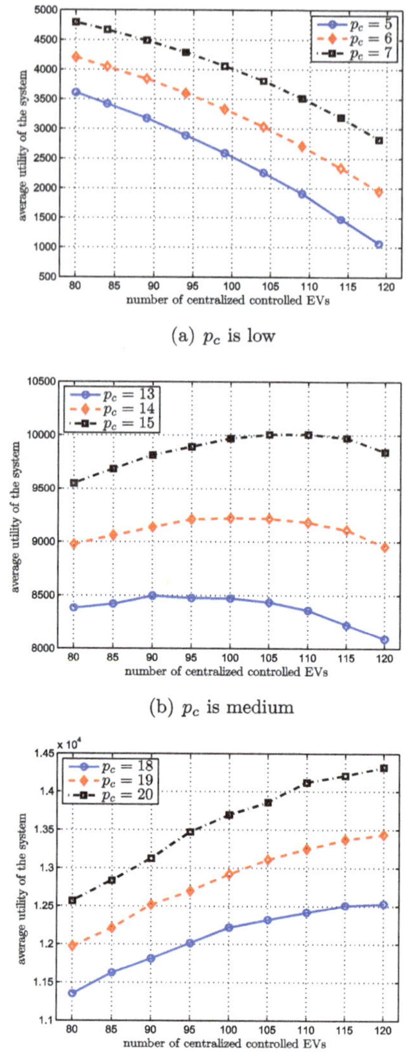

(a) p_c is low

(b) p_c is medium

(c) p_c is high

and we vary the proportion of centralized controlled vehicles (equivalently, the proportion of uncontrolled EVs) and compute the average system utilities. Specifically, the number of centralized controlled vehicle is varied from 80 to 120. We compare the cases where the centralized charging price for one unit of electricity p_c is low, medium, and high. It is shown (see Fig. 7.9a) that when p_c is low, the total system utility drops as the centralized controlled EVs' proportion increases. This fact indicates that at this price stage, the incremental revenue due to the number growth of the centralized controlled EVs is less than the loss caused by the departure of decentralized controlled vehicles. Then, we rise the values of p_c to a medium level and recompute the corresponding average system revenues. We note in Fig. 7.9b that the system utility first increases and then declines as the number of centralized controlled EVs varies from 80 to 120. If p_c is further increased, the system revenue will show a growing tendency as depicted in Fig. 7.9c. This is because at this price interval, the increasing earning from centralized controlled EVs has surpassed the loss due to the decreasing number of decentralized controlled vehicles. We may see that given a price p_c, there exists an optimal number ratio between centralized controlled EVs and decentralized controlled EVs which maximize the system utility. Therefore, based on the EVs' information and cost functions, the system controller can properly choose the energy cap E_m^h, electricity price p_c, and other parameters such that the number ratio of these two EV groups is stimulated to its best value. Proper parameter selections can be obtained by various methods, typically involving large-scale simulations and analyzing a large number of historical data. In addition, given the best number ratio of these two charging groups, the charging park can properly determine the scales of centralized charging facilities and decentralized charging facilities so that the expected revenue is maximized. Hence, our research may provide some illuminations on the investment policy makings of the charging parks.

7.6 Conclusion

In this chapter, we investigate the coordination of EV charging at a charging park considering the EV owners' various charging preferences. A hybrid centralized–decentralized charging mechanism is designed to determine the charging rates and demands of EVs. Specifically, at the centralized charging side, based on the EVs' arrival/departure patterns, a cost minimization problem is formulated and solved to obtain an off-line global optimal scheduling. Considering the fact that the charging station is dynamic with EVs' patterns unpredictable, a model predictive control (MPC)-based adaptive charging approach is developed to determine the near-optimal EV charging profiles in real time. On the decentralized charging side, to model the interactions between EVs and the charging system, a leader–follower noncooperative Stackelberg game-based approach is proposed, where the system controller acts as the leader and EVs act as the followers. We prove the existence and optimality of the equilibrium state. It is also shown that the communication burden between EVs and the system controller is low and our decentralized charging scheme is robust to poor communication channels. Simulation results investigate the performances of the charging scheme and the impacts of different parameters. It is indicated that an optimal charging cap exists for the decentralized charing group which could maximize the revenues of the whole charging system. In addition, our research may further shed some illuminations on the investment policy making for charging park.

References

1. International energy agency: key world energy statistics (2014), https://www.iea.org/publications/freepublications/publication/KeyWorld2014.pdf
2. Y. He, B. Venkatesh, L. Guan, Optimal scheduling for charging and discharging of electric vehicles. IEEE Trans. Smart Grid **3**(3), 1095–1105 (2012)
3. W. Tang, S. Bi, Y.J. Zhang, Online coordinated charging decision algorithm for electric vehicles without future information. IEEE Trans. Smart Grid **5**(6), 2810–2824 (2014)
4. E.F. Camacho, C.B. Alba, *Model Predictive Control* (Springer, London, 2004)
5. F. Allgöwer, A. Zheng, *Nonlinear Model Predictive Control*, vol. 26 (Birkhäuser, Basel, 2000)
6. F. Facchinei, C. Kanzow, Generalized nash equilibrium problems. Ann. Oper. Res. **175**(1), 177–211 (2010)
7. D. Ardagna, B. Panicucci, M. Passacantando, A game theoretic formulation of the service provisioning problem in cloud systems, in *Proceedings of the 20th International Conference on World Wide Web* (ACM, 2011), pp. 177–186
8. A. Nagurney, *Network Economics: A Variational Inequality Approach*, vol. 10 (Springer Science & Business Media, Berlin, 2013)
9. M.V. Solodov, B.F. Svaiter, A new projection method for variational inequality problems. SIAM J. Control Optim. **37**(3), 765–776 (1999)
10. F. Tinti, Numerical solution for pseudomonotone variational inequality problems by extragradient methods, in *Variational Analysis and Applications* (Springer, 2005), pp. 1101–1128
11. W. Tushar, W. Saad, H.V. Poor, D.B. Smith, Economics of electric vehicle charging: a game theoretic approach. IEEE Trans. Smart Grid **3**(4), 1767–1778 (2012)

Chapter 8
Summary and Future Work

In this chapter, we summarize this book and discuss the future work for intelligent control of microgrids and scheduling of EVs' charging.

8.1 Summary of Contributions

In this book, we specifically investigate the intelligent control and scheduling of two important components in smart grid, namely microgrids and EVs. In particular, we first study the energy management problems in microgrids with system uncertainties under different scenarios. We then focus on the control of EV charging to achieve a cost-effective scheduling. The main contributions of this book are summarized as follows.

- In Chap. 3, we propose a novel power demand and supply management scheme in the microgrid to intelligently schedule the energy consumption patterns of home appliances and output of electricity generators. We develop a novel uncertainty model to capture the randomness of renewable energy generation which, by introducing a reference distribution according to past observations and empirical knowledge and defining a distribution set to confine the uncertainty, allows us to conveniently handle the fluctuations of energy supply brought by the renewable energy. We then formulate an optimization problem to determine the optimal power consumption and generation scheduling profiles for minimizing the fuel cost. Finally, we propose a two-stage optimization approach to transform and then solve the prime problem. Numerical results indicate that the proposed scheme help effectively reduce the energy cost. Detailed studies on the impacts of different factors on the proposed scheme provide some interesting insights which shall be useful for policy making for the future MGCC.

© Springer Nature Singapore Pte Ltd. 2018

R. Wang et al., *Intelligent Microgrid Management and EV Control Under Uncertainties in Smart Grid*, https://doi.org/10.1007/978-981-10-4250-8_8

- In Chap. 4, we formulate a cost minimization problem to intelligently schedule energy generations for microgrids equipped with unstable renewable sources and combined heat and power (CHP) generators. In such systems, the fluctuant net demands (i.e., the electricity demands not balanced by renewable energies) and heat demands impose unprecedented challenges. To cope with the fluctuation nature of net demand and heat demand, we develop a new flexible uncertainty model. Specifically, reference distributions are introduced according to predictions and field measurements and then uncertainty sets are defined to confine net and heat demands. The model allows the net demand and heat demand distributions to fluctuate around their reference distributions. Another difficulty existing in this problem is the indeterminate electricity market prices. We develop chance-constraint approximations and robust optimization approaches to firstly transform and then solve the prime problem. We also discuss the extensions of the proposed approaches to handle even wider applications. Numerical results based on real-world data evaluate the impacts of different parameters. It is shown that our energy generation scheduling strategy performs well and the integration of CHP generators can effectively reduce the system expenditure. Our research also helps shed some illuminations on the investment policy making for microgrids.
- In Chap. 5, a cost minimization problem is formulated to intelligently schedule energy generations for microgrids equipped with unstable renewable sources and energy storages. In such systems, the uncertain renewable energy will impose unprecedented scheduling challenges. To cope with the fluctuate nature of the renewable energy, an uncertainty model based on renewable energies' moment statistics is developed. Specifically, we obtain the mean vector and second-order moment matrix according to predictions and field measurements and then define uncertainty set to confine the renewable energy generation. The uncertainty model allows the renewable energy generation distributions to fluctuate within the uncertainty set. We develop chance-constraint approximations and robust optimization approaches based on a Chebyshev inequality framework to firstly transform and then solve the scheduling problem. Numerical results based on real-world data traces evaluate the performance bounds of the proposed scheduling scheme. It is shown that the temporal-correlation information of the renewable energy within a proper time span can effectively reduce the conservativeness of the solution. Moreover, detailed studies on the impacts of different factors on the proposed scheme provide some interesting insights which shall be useful for the policy making for the future microgrids.
- In Chap. 6, we investigate the cost-effective scheduling approach of EV charging at a renewable energy aided charging station. A novel two-stage EV charging mechanism is designed in this chapter. Specifically at the first stage, based on the prediction of future energy requests and considering the elastic charging property of EVs, we formulate an offline optimal energy generation scheduling problem and solve it in a day-ahead manner to determine the energy generation in each time slot next day. Then at the second stage, based on the planned energy generation day-ahead, we develop an adaptive real-time charging strategy to determine the charging rate of each vehicle in a dynamic manner. A charging rate compression

(CRC) algorithm is developed which tremendously reduces the complexity of the problem solving. The fast algorithm supports real-time operations and enables the large-scale small-step scheduling more efficiently. Simulation results indicate that the proposed scheme can help effectively save the energy cost and reduce the system PAR. Detailed evaluations on the impact of renewable energy uncertainties show that our proposed approach achieves a good performance in enhancing the system fault tolerance against uncertainties and the noises of real-time data. We further extend the mechanism to track a given load profile and handle the scenario where EVs only have several discrete charging rates.

- In Chap. 7, we investigate the coordination of EVs' charging at a charging park considering the EV owners' various charging preferences. A hybrid centralized–decentralized (HCD) EV charging control scheme is designed. Specifically on the centralized charging side, we first develop an offline optimal scheduling approach aiming at minimizing the energy cost while satisfying the charging requirements of EVs. Then to deal with the system dynamics and uncertainties, we develop a model predictive control (MPC)-based adaptive scheduling strategy to determine the near-optimal EV charging profiles in real time. On the decentralized side, we model the interactions between EVs and the charging system controller as a leader–follower noncooperative Stackelberg game in which the system controller acts as the leader and the EVs act as followers. The existence of the equilibrium state and its optimality are analyzed. It is shown that the decentralized charging scheme is robust to unstable communication channels. We further investigate the interactions between these two charging groups and propose an algorithm maximizing the total revenues of the whole system. Our research shall provide useful insights helping the charging park operator develop rational investment strategies.

8.2 Future Work

In the future, we will extend the work in the following aspects.

8.2.1 Energy Storage Integration into the Microgrid

Future studies on the microgrid may consider the cases where there is energy storage in the system or surplus energy can be sold to the outside utility grid. For the former case, the energy storage will impose its own cost; meanwhile it may to a certain extent alleviate the uncertainty problem caused by the fluctuation of the renewable energy, especially when the storage is of a large enough capacity. As to the optimal capacity of the energy storage system, it depends on the fluctuation patterns of the energy demand and scale of the microgrid system. For the latter case, a few other conditions, e.g., the price of the electricity, the competition between different microgrids, etc., probably have to be taken into account. The optimization problems for these two different cases therefore become significantly different from the one we considered in this book and worth further studies.

8.2.2 Design of a Vehicle to Grid (V2G) Aggregator

Currently, we only consider EV charging strategies in which EVs obtain electricity from the grid. Actually, EV batteries can also be designed for fast discharge to the grid. System enabling this process is called vehicle to grid (V2G) system. If a group of parked cars with V2G capability are aggregated, they can provide ramping and regulation in support of the power grid. In V2G, one challenge is the availability of EVs, since an EV can only deliver power to the grid when it is parked and connected to grid. As a result, this increases the uncertainty of the power supplied by EVs. How to effectively organize and control these vehicles remains an interesting topic. In the future, we may consider this issue by trying to design an efficient V2G aggregator, which acts as an interface between power grid and EVs. Efficient scheduling algorithms also need to be developed. By properly charging and discharging the batteries of EVs, EVs can act as dispatchable energy storage system, balancing the demand and supply as well as improving the system flexibility and reliability.

8.2.3 More Detailed Statistical Properties of Renewable Energy Generation

For energy generation scheduling problem in microgrids, we may study in more details regarding the statistical properties of renewable energy generation. We may verify that the probability density function of renewable energy generation is unimodal through analyzing a large amount of historical data. If that is the case, we could adopt the generalized Gaussian inequality bounds rather than the generalized Chebyshev inequality bounds to further reduce the conservativeness of the robust solution in Chap. 5. In addition, we may consider a more practical case where the size of the energy storage is limited. Then, the energy generation scheduling strategies may change, and some other problems such as how to determine a proper size of the energy storage device are interesting and worth further studies.

Appendix A
Energy Generation Scheduling in Microgrids

A.1 Proof of Proposition 4.1

Proof Rewrite (4.15)–(4.17) as follows:

$$\max_{f_0(L^h)} \int_0^{+\infty} h(L^h, \mathcal{L}^h) \cdot f_0(L^h) \mathrm{d}L^h \tag{A.1}$$

$$\text{s.t.} \int_0^{+\infty} [\ln f_0(L^h) - \ln g_h(L^h)] f_0(L^h) \mathrm{d}L^h \leq D_h$$

$$\int_0^{+\infty} f_0(L^h) \mathrm{d}L^h = 1.$$

We can see that the objective function and equality constraint function are affined with respect to $f_0(L^h)$. Next, we show that the inequality constraint function is convex.

Lemma: If $f : \mathbf{R^n} \longrightarrow \mathbf{R}$ is convex, then the perspective of f, which is denoted as a function $g : \mathbf{R^{n+1}} \longrightarrow \mathbf{R}$ that

$$g(x, t) = t f(x/t), \tag{A.2}$$

with domain

$$\mathbf{dom}\, g = \{(x, t) | x/t \in \mathbf{dom}\, f,\ t > 0\} \tag{A.3}$$

preserves convexity.

That is to say, if f is a convex function, so is its perspective function g. Similarly, if f is concave, so is g. This can be proved in several ways, e.g., by direct verification of the defining inequality or using epigraphs and the perspective mapping on $\mathbf{R^{n+1}}$. Readers can refer to [1] for more detailed discussions.

© Springer Nature Singapore Pte Ltd. 2018
R. Wang et al., *Intelligent Microgrid Management and EV Control Under Uncertainties in Smart Grid*, https://doi.org/10.1007/978-981-10-4250-8

We consider the convex function $f(x) = -\ln x$ on $\mathbf{R}_{++}$. Its perspective is

$$g(x, t) = -t \ln(x/t) = t \ln(t/x) = t(\ln t - \ln x) \tag{A.4}$$

and it is convex on $\mathbf{R}^2_{++}$. The function g is called the relative entropy of t and x. Then, we have that the KL divergence $\int_{x \in S} [\ln f(x) - \ln g(x)] f(x) dx$ between distribution $f(x)$ and $g(x)$ is convex in $f(x)$ (and $g(x)$ as well). In this case, we claim that the inequality constraint is convex with respect to distribution $f_0(L^h)$. ■

A.2 Reformulation of Problem (4.6)

Specifically, the robust counterpart of Problem (4.6) is as follows:

$$\begin{aligned}
\min_{\mathbf{X},\mathbf{Y},\mathbf{Z},\mathbf{V},\mathbf{U}} \quad & \sum_{h=1}^{H} \Big\{ p_g \cdot U^h + \hat{p}_s^h \cdot V^h + \\
& \sum_{a \in \mathcal{A}} \Big[c_a^m \cdot x_a^h + c_a^b \cdot y_a^h + c_a^s \cdot z_a^h \Big] \Big\} \\
& + \max_{\{W_0 | W_0 \subseteq J_0, |W_0| \leq \Gamma\}} \Big\{ \sum_{h \in W_0} d^h \cdot V^h \Big\} \\
\text{s.t.} \quad & z_a^h \geq 0, \; z_a^h \geq y_a^h - y_a^{h-1} \\
& (4.2)\ (4.3)\ (4.4),\ y_a^h, z_a^h \in \{0, 1\} \\
& x_a^h, V^h, U^h \in \mathbb{R}_0^+, h \in \mathcal{H}, a \in \mathcal{A},
\end{aligned} \tag{A.5}$$

Proposition 4.4: Problem (A.5) has an equivalent MIP formulation as (4.29).

Proof Given a vector $\mathbf{V}^*$, we can convert the last part of Problem (A.5)'s objective function to a linear one as follows:

$$\begin{aligned}
\beta_0(\mathbf{V}^*) &= \max \Big\{ \sum_{h \in W_0} d^h \cdot V^{h*} : W_0 \subseteq J_0, |W_0| \leq \Gamma\} \Big\} \\
&= \max \Big\{ \sum_{h \in J_0} d^h \cdot V^{h*} \cdot \phi_h : \sum_{h \in J_0} \phi_h \leq \Gamma, \\
&\quad 0 \leq \phi_h \leq 1, \forall h \in J_0 \Big\}.
\end{aligned} \tag{A.6}$$

Next, the dual of Problem (A.6) is:

$$\min \sum_{h \in J_0} e^h + \Gamma \cdot \phi \tag{A.7}$$
$$\text{s.t.} \quad \phi + e^h \geq d^h \cdot V^{h*}$$
$$\phi \geq 0, e^h \geq 0, \forall h \in J_0.$$

By strong duality, we have:

$$\beta_0(\mathbf{V}^*) = \min \left\{ \sum_{h \in J_0} e^h + \Gamma \cdot \phi : \right. \tag{A.8}$$
$$\left. \phi + e^h \geq d^h \cdot V^{h*}, \phi \geq 0, e^h \geq 0, \forall h \in J_0 \right\}.$$

Substituting (A.8) to Problem (A.5), we obtain that Problem (A.5) is equivalent to Problem (4.29). ■

Reference

1. S. Boyd, L. Vandenberghe, *Convex Optimization* (Cambridge university press, Cambridge, 2004)

Appendix B
Massive Electric Vehicle Charging Involving Renewable Energy

B.1 Proof for Lemma 5.1

Proof We prove the lemma by adopting the Karush–Kuhn–Tucker (KKT) optimality conditions for the solution to the given problem. The Lagrangian function for Problems (6.7)–(6.10) is:

$$\begin{aligned}
&L(\mathbf{V}, \lambda, \nu) \\
&= \sum_{\tau_i \in \Gamma} w_i (V_{i_{max}} - V_i)^2 + \lambda_0 \left(\sum_{\tau_i \in \Gamma} V_i - V_m \right) \\
&+ \sum_{\tau_i \in \Gamma} \lambda_i (V_{i_{min}} - V_i) + \sum_{\tau_i \in \Gamma} \nu_i (V_i - V_{i_{max}}),
\end{aligned} \tag{B.1}$$

where $\lambda_0 \geq 0$, $\lambda_i \geq 0$, and $\nu_i \geq 0$ for $\tau_i \in \Gamma$ are Lagrangian multipliers associated with constraints (6.8), (6.9), and (6.10). Through Slater's condition, strong duality holds for this problem. In such case, the sufficient and necessary conditions for the existence of a minimum value at V_i^* are, for all $\tau_i \in \Gamma$

$$\frac{\partial L}{\partial V_i^*} = -2w_i (V_{i_{max}} - V_i^*) + \lambda_0 - \lambda_i + \nu_i = 0, \tag{B.2}$$

$$\lambda_0 \left(\sum_{\tau_i \in \Gamma} V_i - V_m(t) \right) = 0, \tag{B.3}$$

$$\sum_{\tau_i \in \Gamma} \lambda_i (V_{i_{min}} - V_i) = 0, \tag{B.4}$$

$$\sum_{\tau_i \in \Gamma} \nu_i (V_i - V_{i_{max}}) = 0. \tag{B.5}$$

© Springer Nature Singapore Pte Ltd. 2018

R. Wang et al., *Intelligent Microgrid Management and EV Control Under Uncertainties in Smart Grid*, https://doi.org/10.1007/978-981-10-4250-8

First assume that (6.8) is inactive, which means that $\sum_{\tau_i \in \Gamma} V_i^* - V_m(t) < 0$ and $\lambda_0 = 0$. In this case, at least one constraint in (6.9) or (6.10) must be active. Let us assume that the kth constraint in (6.9) is active, i.e., $V_k^* = V_{k_{min}}$ and $\lambda_k \geq 0$. Then, the kth constraint in (6.10) must be inactive, that is, $V_k^* - V_{k_{max}} < 0$ and $\nu_k = 0$. From (B.2), we then obtain

$$\lambda_k = -2w_i(V_{i_{max}} - V_i^*) < 0, \tag{B.6}$$

which contradicts the assumption that $\lambda_k \geq 0$. Hence, we have the conclusion that if any $V_k^* = V_{k_{min}}$, constraint (6.8) have to be active.

Similarly, if at least one constraint in (6.10) is active while others are inactive, i.e., $V_h^* = V_{h_{max}}$ (active) and $V_{l_{min}} < V_l^* < V_{l_{max}}$ (inactive), then we can obtain that $\lambda_h = 0$, $\nu_h \geq 0$, $\lambda_k = 0$, and $\nu_k = 0$. Based on (B.2), we obtain the following two equations:

$$\lambda_0 = 2w_i(V_{h_{max}} - V_h^*) + \lambda_h - \nu_h = -\nu_h \leq 0, \tag{B.7}$$

$$\begin{aligned} \lambda_0 &= 2w_i(V_{k_{max}} - V_k^*) + \lambda_k - \nu_k \\ &= 2w_i(V_{k_{max}} - V_k^*) > 0. \end{aligned} \tag{B.8}$$

Note that the above equations (B.7) and (B.8) cannot be satisfied simultaneously, which means that all the constraints in (6.10) can either be active or inactive. Under such cases, if all the constraints in (6.10) are active, we have

$$\sum_{\tau_i \in \Gamma} V_i^* = \sum_{\tau_i \in \Gamma} V_{i_{max}} > V_m(t), \tag{B.9}$$

which contradicts the constraint (6.8) that the charging task is schedulable. If all the constraints in (6.10) are inactive, then from (B.5) we have $\lambda_0 = 0$ from (B.8), which means that $\sum_{i=1}^{M(t)} V_i^* - V_m(t) = 0$ given (B.3). This again contradicts the assumption that (6.8) is inactive. Therefore, we have the conclusion that for any solution to the optimization problem (6.7)–(6.10), constraint (6.8) is active, i.e., $\sum_{\tau_i \in \Gamma} V_i^*(t) = V_m(t)$ and $V_i^*(t) \neq V_{i_{max}}$, for $\tau_i \in \Gamma$. Hence, Lemma 5.1 is proved. ■

B.2 Proof for Theorem 5.1

Proof Consider the KKT optimality condition in (B.2)–(B.5). We have proved in Lemma 5.1 that any solution, $V_i^*(t)$, to the optimization problem must satisfy $\sum_{\tau_i \in \Gamma} V_i^*(t) = V_m(t)$ and $V_i^*(t) \neq V_{i_{max}}$, for $\tau_i \in \Gamma$. Therefore, we only need to consider the condition that $\nu_i = 0$, for $\tau_i \in \Gamma$. Suppose that the hth constraint in (6.9) is active, i.e., $V_h^* = V_{h_{min}}$ and

$$\begin{aligned}\lambda_h &= \lambda_0 + \nu_0 - 2w_h(V_{h_{max}} - V_h^*) \\ &= \lambda_0 - 2w_h(V_{h_{max}} - V_{h_{min}}).\end{aligned} \tag{B.10}$$

For other constraints that are inactive, we have $\lambda_k = 0$ based on (B.4). Based on (B.2), we have:

$$\frac{\lambda_0}{w_i} = \frac{\lambda_i}{w_i} + 2(V_{i_{max}} - V_i^*). \tag{B.11}$$

By summing up the above equation for all i that satisfy $V_i^* \neq V_{i_{min}}$, we can get:

$$\lambda_0 \sum_{V_i^* \neq V_{i_{min}}} \frac{1}{w_i} = 2 \sum_{V_i^* \neq V_{i_{min}}} (V_{i_{max}} - V_i^*), \tag{B.12}$$

which is equivalent to

$$\begin{aligned}&\lambda_0 \sum_{V_i^* \neq V_{i_{min}}} \frac{1}{w_i} \\ &= 2\Bigg(\sum_{V_i^* \neq V_{i_{min}}} V_{i_{max}} + \sum_{V_i^* = V_{i_{min}}} V_{i_{min}} \\ &\quad - \sum_{V_i^* = V_{i_{min}}} V_{i_{min}} - \sum_{V_i^* \neq V_{i_{min}}} V_i^* \Bigg) \\ &= 2\left(\widehat{V}(t) - V_m(t)\right),\end{aligned} \tag{B.13}$$

and thus:

$$\lambda_0 = \frac{2\left(\widehat{V}(t) - V_m(t)\right)}{\sum_{V_i^* \neq V_{i_{min}}} (1/w_i)} \tag{B.14}$$

as long as $\widehat{V}(t) > V_m(t)$, $\lambda_0 > 0$, $\lambda_i \geq 0$, and constraint (6.10) are satisfied. Under such case, the optimal charging rate V_i^* either satisfies $V_i^* = V_{i_{min}}$ or

$$\begin{aligned}V_i^* &= V_{i_{max}} - \frac{\lambda_0}{2w_i} \\ &= V_{i_{max}} - \frac{\frac{1}{w_i}\left(\widehat{V}(t) - V_m(t)\right)}{\sum_{V_i^* \neq V_{i_{min}}} (1/w_i)}.\end{aligned} \tag{B.15}$$

Since Slater condition holds for Problems (6.7)–(6.10), the KKT conditions provide necessary and sufficient condition for optimality. Theorem 5.1 is proven. ■

Appendix C
Hybrid Charging Control of Electric Vehicles

C.1 Proof of Theorem 6.3

Proof Suppose that $\mathbf{x}^1$ and $\mathbf{x}^*$ are both solutions and $\mathbf{x}^1 \neq \mathbf{x}^*$. Then since both $\mathbf{x}^1$ and $\mathbf{x}^*$ are solutions, they must satisfy:

$$\mathbf{F}(\mathbf{x}^1)^T \cdot (\mathbf{x}^{'} - \mathbf{x}^1) \geq 0, \ \forall \mathbf{x}^{'} \in \mathbf{X}, \tag{C.1}$$

$$\mathbf{F}(\mathbf{x}^*)^T \cdot (\mathbf{x}^{'} - \mathbf{x}^*) \geq 0, \ \forall \mathbf{x}^{'} \in \mathbf{X}. \tag{C.2}$$

After substituting $\mathbf{x}^*$ for $\mathbf{x}^{'}$ in (C.1) and $\mathbf{x}^1$ for $\mathbf{x}^{'}$ in (C.2) and adding the resulting inequalities, we obtain:

$$\mathbf{F}(\mathbf{x}^1 - \mathbf{x}^*)^T \cdot (\mathbf{x}^* - \mathbf{x}^1) \geq 0. \tag{C.3}$$

But inequality (C.3) is in contradiction to the definition of strict monotonicity. Hence, $\mathbf{x}^1 = \mathbf{x}^*$. ■

C.2 Proof of Theorem 6.4

Proof Given a fixed electricity price p_d^h, to find the socially optimal solution of the proposed followers' GNEP, one has to solve the following optimization problem:

$$\max_{x_b^h} \sum_{b \in \mathcal{B}(h)} U_b^h(x_b^h, \mathbf{x}_{-b}^h, \alpha_b^h, \beta_b^h, p_d^h) \tag{C.4}$$

$$\text{s.t.} \sum_{b \in \mathcal{B}(h)} x_b^h \leq E_m^h, \tag{C.5}$$

© Springer Nature Singapore Pte Ltd. 2018
R. Wang et al., *Intelligent Microgrid Management and EV Control Under Uncertainties in Smart Grid*, https://doi.org/10.1007/978-981-10-4250-8

which is obviously a quadratic programming problem. The Karush–Kuhn–Tucker (KKT) optimal conditions for this problem are:

$$\mathbf{F}^h + \nabla_{\mathbf{x}}^h \left(\sum_{b \in \mathcal{B}(h)} x_b^h - E_m^h \right) \cdot \lambda = 0, \tag{C.6}$$

$$\lambda \left(\sum_{b \in \mathcal{B}(h)} x_b^h - E_m^h \right) = 0, \tag{C.7}$$

which are exactly the same to the KKT conditions of the VI($\mathbf{X}^h$, $\mathbf{F}^h$) problem, i.e., (7.26) and (7.27). Since the Slater's condition holds, the KKT conditions provide sufficient and necessary conditions for optimality. Thus, the unique VE is the socially optimal solution of the proposed followers' GNEP. ■

FSC
www.fsc.org
MIX
Papier | Fördert
gute Waldnutzung
FSC® C083411

Kolibri 360 GmbH
Ferdinand-Jühlke-Straße 7
99095 Erfurt, Deutschland
produktsicherheit@kolibri360.de